Niche Tourism and Sustainability: Perspectives, Practices and Prospects

Niche Tourism and Sustainability: Perspectives, Practices and Prospects

Edited by

Anna Farmaki
Cyprus University of Technology, Limassol, Cyprus

Pramendra Singh
Lovely Professional University, Punjab, India

and

Viana Hassan
American University of Malta, Cospicua, Malta
CEO and Founder of European Tourism & Events Training Academy

CABI

CABI is a trading name of CAB International

CABI
Nosworthy Way
Wallingford
Oxfordshire OX10 8DE
UK

Tel: +44 (0)1491 832111
E-mail: info@cabi.org
Website: www.cabi.org

CABI
200 Portland Street
Boston
MA 02114
USA

T: +1 (617)682-9015
E-mail: cabi-nao@cabi.org

A catalogue record for this book is available from the British Library, London, UK.

Library of Congress Cataloging-in-Publication Data

ISBN-13: 9781800626645 (hardback)
9781800626652 (ePDF)
9781800626669 (ePub)

DOI: 10.1079/9781800626669.0000

Commissioning Editor: Rebecca Stubbs
Editorial Assistant: Emma McCann
Production Editor: Shankari Wilford

Typeset by Exeter Premedia Services Pvt Ltd, Chennai, India
Printed and bound in the UK by CPI Group (UK) Ltd, Croydon, CR0 4YY

Contents

Biographies of Editors and Contributors

Editors

Anna Farmaki (ORCID: 0000-0002-9996-5632) is Assistant Professor at the Cyprus University of Technology, Cyprus. She holds a doctorate (PhD) in marketing from Nottingham Trent University, UK. She has published extensively in reputable, peer-reviewed academic journals such as *Tourism Management*, *Annals of Tourism Research*, *Journal of Travel Research*, *Journal of Business Research*, *Journal of Sustainable Tourism* and *International Journal of Hospitality Management*, among others, and has presented her work in various international conferences, seminars and workshops. Dr Farmaki is a Fellow of the Higher Education Academy (UK) and a member of the Management Committee in several COST Actions. In addition, she is Associate Editor of *Tourism Review* and sits on the Editorial Board of various journals including *International Journal of Contemporary Hospitality Management*, *Service Industries Journal*, *Tourism Planning & Development*, *Journal of Hospitality and Tourism Management* and *Journal of Service Theory and Practice*. Her research interests lie primarily in the areas of tourism planning and development, with emphasis on sustainable tourism, and tourist behaviour. Email: anna.farmaki@cut.ac.cy

Pramendra Singh (ORCID: 0000-0002-9142-265X) is currently working as an Assistant Professor and Head of the Department of Tourism and Airlines in the School of Hotel Management and Tourism at Lovely Professional University, Punjab, India. He completed his PhD in heritage tourism. He has also earned three master's degrees, two in Tourism and Travel Management and one in History, along with a Postgraduate Diploma in Folklore and Culture Studies. He has more than 10 years of teaching and research experience. His research portfolio includes many research papers, book chapters and edited books on tourism and related areas published by internationally reputed publishers. His research interests include heritage tourism, cultural tourism, sustainable tourism, niche tourism, tourism education, and tourism and technology.
Email: pramendra58@gmail.com

Viana Hassan (ORCID: 0000-0002-7372-5059) is actively involved in academic and research work in Tourism and Cultural Management in Lebanon and Malta. She is the Founder and CEO of the European Tourism & Events Training Academy. She received her PhD in Tourism Management and Culture from Saint Joseph University, Beirut. Her thesis was entitled 'Medical Tourism in Lebanon'. She is an experienced lecturer with more than 16 years of teaching experience at several universities

in Lebanon (Lebanese University, AUL, Lebanese International University, Islamic University, the Institute of Tourism Studies, STC higher education partner of Wolverhampton University) and in Malta (University of Malta, GBS Malta, Scholars School System). She is interested in tourism management, customer service, branding, event management, airline business management, niche tourism, business operations management, customer acquisition and medical tourism. Dr Hassan has participated in several international conferences (e.g. Expo Dubai 2020, 7th Medical Tourism World Summit).
Email: viana24@hotmail.com

Contributors

Maria Alebaki holds a PhD in Agricultural Economics from the Aristotle University, Greece, supported by the National Scholarships Foundation. In addition, she has conducted two postdoctoral fellowships at the Hellenic Agricultural Organization DEMETER (ELGO-DIMITRA) and the Agricultural University of Athens. Since 2018, she is an Assistant Researcher at the Agricultural Economics Research Institute of ELGO-DIMITRA (AGRERI). Apart from her position at AGRERI, she serves as a Visiting Professor (Agricultural University of Athens, Hellenic Mediterranean University) and an Adjunct Faculty Member (Hellenic Open University). Her current research interests include the exploration of sustainability, resilience, training aspects and digital marketing of wine tourism.
Email: malempaki@gmail.com

Katerina Antoniou is an interdisciplinary scholar in Tourism and International Relations and an Assistant Professor and Course Leader at the University of Central Lancashire Cyprus (UCLan Cyprus). She has published work on peace tourism, citizen diplomacy and social mediation, and is the author of the book *Tourism as a Form of International Relations*. She is a resident expert for the Interdisciplinary Centre for Law, Alternative and Innovative Methods (ICLAIM). Dr Antoniou is a member of the initiative 'Social Mediation in Practice' (European Citizen's Prize 2020) and has co-authored a series of Social Mediation manuals. She is a Fulbright Alumna.
Email: kaantoniou@uclan.ac.uk

Kleio Arvanakou holds a bachelor's degree in Business Administration of Food and Agricultural Enterprises from the University of Patras (Greece) and a master's degree in Tourism Management from the Hellenic Open University (Greece). Her thesis, titled 'Adopting the Dynamic Capabilities Approach to Explore Sustainable Wine Tourism in Corfu Island', focused on advancing sustainable wine tourism through the lens of dynamic capabilities. She has extensive experience as a Front Office Manager in luxury hotels, specializing in optimizing guest experiences. Throughout her career, she has successfully implemented strategies to enhance guest satisfaction and develop innovative solutions to ensure smooth departmental operations.
Email: kleiwarvanak@gmail.com

Furkan Atasoy Karacabey was born in 1996 in Yozgat, Türkiye. He works as a Research Assistant in the Department of Tourism Management at the Faculty of Tourism, Sivas Cumhuriyet University. He is pursuing his doctoral studies at Ankara Hacı Bayram Veli University, and he has completed his master's and bachelor's degrees in Tourism Management at İzmir Katip Çelebi University. His research areas include tourism management and residents' support for tourism development. Karacabey has conducted various academic studies in the fields of tourism and hotel management.
Email: furkankaracabey@cumhuriyet.edu.tr

Bruno Barbosa de Sousa is a distinguished academic with a PhD in Marketing and Strategy. He graduated in Business Management in 2006 and has since held various academic positions.

Currently, he serves as a Lecturer at the Polytechnic Institute of Cávado and Ave (IPCA, Portugal) and has previously lectured at the University of Minho. Professor Sousa's research focuses on diverse areas including sustainable development, tourism behaviour, organizational performance and agile marketing. He has authored several books and numerous book chapters, along with contributing to prestigious journals such as *Current Issues in Tourism, EuroMed Journal of Business, Annals of Leisure Research, Journal of Cultural Heritage Management and Sustainable Development, Tourism Management Perspectives, Wine Economics and Policy, Journal of Hospitality and Tourism Insights, Journal of Tourism, Heritage & Services Marketing* and *Journal of Organizational Change Management.*
Email: bsousa@ipca.pt

Gülsel Çiftci is an Associate Professor at Trakya University's Faculty of Applied Sciences in Edirne, Türkiye. She is also a Visiting Research Scholar at the School of Community Resources and Development, Arizona State University, USA. She received her PhD in Business Management and holds a master's degree in Tourism Management. With extensive experience in cross-border tourism, crisis management in the hospitality industry and various research projects, she has been honoured with multiple academic awards. Additionally, she is a licensed tourist guide in Türkiye.
Email: gulsellciftci@gmail.com

Fani Efthymiadou is an Assistant Professor at the Department of Tourism Administration, Faculty of Managerial Sciences of the Boğaziçi University, Istanbul, Türkiye. Her research focuses on the sharing economy, women empowerment and sustainability.
Email: fani.efthymiadou@bogazici.edu.tr

Emrullah Erul, PhD, is Associate Professor in Tourism Management at İzmir Katip Çelebi University, Türkiye. He undertakes community-centred research focusing on sustainable tourism planning and management (with a particular emphasis on social–cultural impacts and resident–tourist relationships) as well as protected areas and adjacent communities. His research interests concern residents' attitudes towards tourism development and their support for tourism, sustainable tourism development and Artificial Intelligence (AI) (e.g. ChatGPT and Chatbots) in tourism and hospitality. His research has appeared in numerous leading tourism journals, such as *Tourism Management, Journal of Sustainable Tourism, Journal of Travel Research* and *Current Issues in Tourism.*
Email: emrullah.erul@ikc.edu.tr

Ângela Gonçalves is pursuing a master's degree in Tourism Management at the Polytechnic Institute of Cávado and Ave (IPCA, Portugal). She graduated in Tourism from the Polytechnic Institute of Viana do Castelo (IPVC) in 2023. During her studies, she participated in a Galileo course, an INPEC+ project and volunteered at an event organized by the Administration of the Ports of Douro, Leixões and Viana do Castelo (APDL). She was also involved in the 'DEMOLA' research project and a module on projects, which ignited her interest in the investigation field. Ângela is currently a research fellow at the IPVC, under the FISATUR project funded by the European Maritime, Fisheries and Aquaculture Fund (EMFAF).
Email: a30004@alunos.ipca.pt

Werner Gronau holds the position as Professor for Tourism, Travel & Transport at the German University of Applied Sciences Stralsund. He is member of research groups such as NECTAR or the German 'Tourism Research Working Group'. He works as Associate Editor of the *Journal of Tourism Science* and *Frontiers in Sustainable Tourism*. His research interests focus on sustainable destination management and tourism-related transport issues. He has worked in several research projects granted by the German Ministry of Research or the European Commission and presented the results at international conferences as well as in various journals and books.
Email: werner.gronau@hochschule-stralsund.de

Jana Große Hokamp earned her BA in Human Geography at the Catholic University of Eichstätt-Ingolstadt. With a master thesis on 'Motorcycle tourism in Tirol' she completed her master studies at the University of Applied Sciences in Stralsund. Since October 2021 she is working as a tourism consultant at 'Kohl und Partner' Stuttgart with focus on destination consultation, regional planning, moderations of workshops, as well as accompanying participatory processes.
Email: jana@gr-hokamp.de

Maria Hadjielia Drotarova received her PhD from the Department of Educational Research of Lancaster University, UK. She is a Head of Research and Assistant Professor in the field of business and hospitality, CTL Eurocollege, Cyprus. She is a UNESCO Peace Speaker. Her research is found at the intersection of education and tourism, drawing on longitudinal qualitative research approach and focusing on topics such as educational research, social learning, dark tourism and sharing economy. Dr Hadjielia has substantial experience on EU-funded projects under Erasmus+ and Horizon 2020.
Email: hadjieliadrm@ctleuro.ac.cy

Sıla Karacaoğlu is an Associate Professor in the Faculty of Applied Sciences, Department of Tourism Guiding at Bilecik Şeyh Edebali University, Türkiye. She received her PhD degree from the Mersin University in Tourism Management in 2017. In her doctoral dissertation, she studied 'Community Perceptions, Attitudes and Support for Community Based Tourism: The Case of Misi Village'. Her research interests are tourism marketing, sustainable tourism and its types, cultural heritage and tourist guiding.
Email: sila.karacaoglu@bilecik.edu.tr

Stella Kladou (PhD, MBA, FHEA) is an Assistant Professor at the Hellenic Mediterranean University (Greece). Her academic experience first developed in the UK (Sheffield Hallam University), Türkiye (Boğaziçi University) and their international partners. Stella's research revolves around place branding and specifically the heritage–culture and tourism dimensions of the place brand. She has co-edited books, published in prominent academic journals and presented at well-esteemed international conferences. She is an International Place Brand Specialist for *The Place Brand Observer*, the Associate Editor for Europe and Africa Region of the *Journal of Foodservice Business Research*, on the Editorial Review Board of the *Annals of Tourism Research*, on the Editorial Board of the *Journal of Qualitative Research in Tourism* and participates in the organizing and/or scientific board of international conferences.
Email: stellakladou@hmu.gr

Andrew Le is a Senior Lecturer at the International College of Management, Sydney (ICMS) in Australia. He earned his PhD in Tourism and Hospitality Management from Griffith University (Australia) and an MSc in Leisure and Environments from Wageningen University (the Netherlands). Dr Le's research focuses on the tourism and hospitality industry and sustainable development. He has published works in top-tier journals and with renowned publishers, including the *Journal of Sustainable Tourism*, Springer, CABI, IGI Global and IUCN. Beyond academia, Andrew has more than 20 years of experience as a Senior Consultant for international development projects. He has worked on initiatives funded by various organizations, including EuropeAid, the Asian Development Bank (ADB), UNESCO, the Netherlands Development Organisation (SNV), the United States Agency for International Development (USAID) and the Australian Agency for International Development (AusAID).
Email: ale@icms.edu.au

Glenn Murray lives and works on Sydney's Northern Beaches and serves as the Program Manager for International Tourism and Hospitality at the International College of Management, Sydney (ICMS). He holds a master's in Research and a master's in International Hospitality and Hotel Management.

Glenn has lectured at several universities in tourism and hospitality management, with a strong emphasis on the experience economy, focusing on both guest and workforce perspectives, including innovation and emerging technologies. His passion for discovering the ultimate experience drives his ongoing travels around the world.
Email: gmurray@icms.edu.au

Athina Nella serves as an Assistant Professor of Tourism Marketing at the School of Social Sciences, Hellenic Open University, Greece. In parallel with her academic pursuits and extensive teaching experience, she amassed 20 years of professional experience in FMCG, Services & Tourism Marketing, enriched by roles within esteemed organizations of the private and public sector in Greece and the UK. Her research interests encompass the domains of tourism marketing, destination branding and tourism behaviour. In 2021 she was honoured with a Fulbright Scholarship (Chico, California State University).
Email: a.nella@eap.gr

Beverly M. Ntshabeleng is a Lecturer at the Tshwane University of Technology (TUT) in South Africa. She obtained her PhD in Management Studies with a specialization in Tourism from the University of South Africa. Her main research areas are sustainable tourism, heritage tourism and visitor behaviour.
Email: NtshabelengBM@tut.ac.za

Feras Orekat is a Lecturer at the International College of Management, Sydney (ICMS) in Australia. He earned his PhD in Heritage Studies from the University of Sydney. Dr Orekat's research focuses on visitors' experiences, heritage tourism and tourism sustainable development. Feras has more than 10 years of experience working as a tour guide and heritage site consultant. He is a developer of innovative tour designs for heritage sites, focusing on sustainability and visitor engagement. He has worked on initiatives funded by various organizations including the University of Sydney, The University of Technology Sydney, The University of Jordan, UNESCO and the Hashemite University.
Email: forekat@icms.edu.au

Chryso Panayidou is an accomplished academic in the field of Tourism Studies. She holds a PhD in Tourism Development from the Cyprus University of Technology, an MA in European Leisure Studies from Loughborough University, UK and Tilburg University, the Netherlands, and a BA (Hons) in Business Studies with Tourism from the University of the West of England, Bristol, UK. Her research interests include tourism development and special interest tourism. She is currently a Specialist Scientist in the Department of Tourism Management and Hospitality at the Cyprus University of Technology.
Email: chryso.panayidou@cut.ac.cy

Vasilis Papavasiliou received his PhD from the School of Hotel and Tourism Management (SHTM) at the Hong Kong Polytechnic University (PolyU). He is the Head of the hospitality department and Lecturer in the field of hospitality and tourism, CTL Eurocollege, Cyprus. He has a keen interest on topics in relation to culture and identity creation. Specifically, his research focuses on understanding how individuals interact independently but also as a collective within the tourism and hospitality world. Hence, Dr Papavasiliou's academic and research interests are cultural anthropology, identity politics, authenticity and tourism destination development.
Email: vpapavasiliou@ctleuro.ac.cy

Mariana Pereira is a student of Tourism Management at the Polytechnic Institute of Cávado and Ave (IPCA, Portugal). In 2023, she graduated in Tourism from the Polytechnic Institute of Viana do Castelo (IPVC), at the School of Technology and Management. During her academic journey,

she participated in projects like 'DEMOLA', studying sustainable food, and attended several lectures, including the Tourism Days. She also achieved a B1 level in English, certified by the Cambridge School. Her academic involvement reflects a strong interest in exploring and contributing to the development of tourism, focusing on sustainable practices and innovation.
Email: a30323@alunos.ipca.pt

Madiseng M. Phori is a tourism lecturer at Tshwane University of Technology (TUT) and holds degrees of doctor of Management Sciences in Tourism from TUT and a master of Management Sciences in Hospitality and Tourism from Durban University of Technology in South Africa. His research focus includes community-based tourism, sustainable tourism development, heritage tourism, rural tourism, pilgrimage tourism as well as safety and security in tourism.
Email: PhoriMM@tut.ac.za

Jéssica Ribeiro is a student at the Polytechnic Institute of Cávado and Ave (IPCA, Portugal), where she is studying for a masters in Tourism Management. She completed her degree in Tourism in 2023 at the Polytechnic Institute of Viana do Castelo (IPVC). During the summer of 2023, she did an internship at a renowned Portuguese travel agency. During her degree, she undertook Galileo training and also volunteered at an event organized by the Douro, Leixões and Viana do Castelo Port Administration (APDL), acting as a tourist guide. With a special interest in research, during her degree she also took part in the 'DEMOLA' research project. She is currently enrolled on the Specialist Accounting Technician course, part of the 'ATIVAR+' initiative run by the Institute for Employment and Vocational Training (IEFP).
Email: a30410@alunos.ipca.pt

Knut Scherhag is Professor of Destination Management at Worms University of Applied Sciences in Germany; Head of the Tourism bachelor's degree programmes (Department of Tourism/Travel Management); representative for the Department's international cooperation; and Deputy Speaker of the Commission Wine Tourism/Culinary of the German Society for Tourism Science (DGT). After studying business administration at the University of Trier, he worked for 4 years as a research assistant at TU Dresden, then for about 5 years as a senior consultant at Europäisches Tourismus Institut GmbH/Trier (ETI). Dr Scherhag's research focus is on special interest tourism in the destination (e.g. motorcycle tourism and wine tourism) and regional tourism development.
Email: scherhag@HS-Worms.de

Dimitrios P. Stergiou holds a PhD from the University of Surrey, UK. Currently he is Vice-Dean and Associate Professor of Tourism Management at the School of Social Sciences, Hellenic Open University, Greece. He has over 15 years of experience in education and research in the tourism field in Greece and internationally. Dimitrios' research interests are in the wider areas of tourism and hospitality management, and theoretical/methodological issues in social (tourism) research. Alongside his academic duties he has worked with several government and private organizations on tourism project work.
Email: dstergiou@eap.gr

Acknowledgements

The editors would like to thank several people who have contributed to the development of this book. First and foremost, we would like to thank all the authors who have trusted us with their chapters. We appreciate the time spent to write the chapters, as well as their understanding and patience throughout the revision process until the publication of the book. Also, we would like to thank Lucy Pritchard, our managing editor, for her invaluable assistance throughout the development of this book. Last, we are grateful to the staff at CABI for their support from the preparation of the book proposal to the submission of the final book content. We would especially like to thank Claire Parfitt and Rebecca Stubbs for their continuous guidance.

Introduction

Anna Farmaki[1]* and Pramendra Singh[2]
[1]Cyprus University of Technology, Limassol, Cyprus; [2]Lovely Professional University, Punjab, India

The Rise of Niche Tourism

Over the years, niche tourism emerged as a popular developmental option for destinations and a preference of tourists. Currently regarded as one of the fastest-growing sectors of the tourism industry (Bunghez, 2021), niche tourism is forecasted to continue to expand in both demand and supply. The popularity of niche tourism is somewhat unsurprising considering its promising potential as a developmental strategy for emerging and established destinations as well as an experience offered to tourists. To begin with, as the global tourism market has become highly competitive, destination management organizations (DMOs) are striving to identify new tourism products that will differentiate and/or improve their offering. Niche tourism has proven to be particularly favourable as a tourism development approach in emerging destinations and peripheral regions (Novelli, 2018) that often have limited resources and are better able to serve small tourist segments. Additionally, niche tourism has been heralded as a preferred rejuvenation strategy for established destinations wishing to diversify from the conventional tourism product as it allows them to expand their product offering and enhance their destination image (Farmaki, 2012a).

Notwithstanding, the development of niche tourism represents a turning point in contemporary tourism that largely evolved as a response to consumer needs (Novelli, 2005). Indeed, there has been a noticeable change in consumer behaviours in the last three decades as a result of globalization and a higher disposable income, highlighting the need for destinations to seek new tourism products that differentiate them and offer unique experiences to an increasingly demanding breed of tourists. According to Lew (2008), niche tourism is compatible to postmodern society characteristics where hyper-consumerism entails that people are exposed to a huge range of products and services. In this context, market differentiation and specialization has become prevalent for tourism businesses and destinations which are called to serve an array of market segments, each possessing distinct features and preferences. Specifically, the postmodern tourist is sophisticated and seeks authentic memorable experiences as well as activities that reinforce his/her identity (McCabe, 2015). Activities that tourists perform at home during their free time and allow people to express their identity (e.g. photography, cycling) are essentially packaged and sold as a holiday option under the label niche tourism. Hence, niche tourism is often referred to as special interest tourism as it targets small groups of people who share a special interest.

This parallel use of terminology creates semantic issues, calling for greater attention to be given to the conceptualization of niche tourism in order to understand what it is, what aspects it

*Corresponding author: anna.farmaki@cut.ac.cy

encompasses and, most importantly, how it distinguishes itself from other forms of tourism that have become prevalent as keywords in recent tourism scholarship and policy.

Definition and Conceptualization of Niche Tourism

The term *niche* is attributed to Hutchinson who first mentioned it in 1957 to describe a region in a large geographical area which a creature can exploit. Since, the word niche has been used in marketing as part of market segmentation strategies adopted by companies to delineate 'a specialised space in the market where a firm can escape direct competition or exploit its own special competence' (Richards, 2021, p. 1). In tourism, the term niche came to be used in reference to a tourism product that meets the needs of a small market segment which includes individuals with shared characteristics. In the last three decades, as the need for product diversification became imminent, destinations' developmental focus shifted from mass tourism to smaller niche tourism segments in an attempt to withstand increasing competition amidst changing consumer needs. None the less, an overview of various conceptualization attempts reveals that niche tourism has to do more with what tourists are doing rather than their number (Bunghez, 2021). The geographical definitional approach, for instance, considers the location of the tourism consumption (e.g. urban, rural, etc.) and the population involved in the process but what matters most is the experience and/or activity that tourists wish to engage in (Novelli, 2005). The product approach includes the activities and amenities offered as part of niche tourism and which influence tourist expectations, whereas the customer approach considers tourist needs and wants as the basis of the experience of niche tourism (Novelli, 2005). Hence, niches have been linked with specific special interests of tourists (Richards, 2021).

Overall, a broad range of niche tourism forms exists based on the interests of tourists. According to Papathanassis (2011), these special interests often reflect societal trends on which niche tourism can be classified, including: (i) sustainability (e.g. ecotourism, agritourism); (ii) the experience economy (e.g. extreme sports); (iii) self-development (e.g. cultural tourism, educational tourism); (iv) hedonism (e.g. dark tourism, sex tourism); (v) consumerism (e.g. shopping tourism); and (vi) conscious living (e.g. religious tourism, medical tourism). Yet, these niche tourism products often vary in characteristics and size. Novelli *et al.* (2022) argued that there are macro niche tourism products that include various activities such as sport tourism and nature tourism among others. Within these, micro niches can be detected that encompass more exclusive activities; for example, under sport tourism activities like trekking, cycling or extreme sports may be commodified and sold as niche tourism products whereas under nature tourism specific forms such as ecotourism or wildlife tourism may be offered.

Therefore, niche tourism involves an extensive list of tourism activities that can appeal to the mainstream tourist as well as more specific groups of tourists who travel to satisfy a specific need. This characteristic of niche tourism, coupled with its growth in popularity over the years, raises questions over the nature of niche tourism in relation to mass tourism. In contrast to mass tourism's standardized products and broad appeal, niche tourism targets a selective audience seeking exclusive and authentic experiences (Marson, 2011). Hence, niche tourism is generally positioned at the antipode of mass tourism (Bunghez, 2021); however, elements between mass tourism and niche tourism frequently overlap. For example, the participation of tourists in the niche tourism experience may not necessarily be active as tourists may engage in the tourism consumption in a passive way (Farmaki, 2012b) with their interaction with locals remaining superficial. Likewise, many large tour operators nowadays offer niche tourism products (Wilhelm Stanis and Barbieri, 2013) entailing that the production patterns of niche tourism resemble those of mass tourism.

Conceptualization attempts are further complexified by associations of niche tourism to the alternative tourism paradigm (Skinner and Soomers, 2019) as both aim at addressing the broader socio-economic concerns arising from mass tourism activities. As such, niche tourism has come to be considered as more sustainable (Trunfio *et al.*, 2006; Boukas and Chourides, 2016) based on the premise that its impacts on the local environment and society are minimal given the small number of tourists involved and the focus on social and environmental aspects. Even so the contribution of niche tourism to sustainability has been contested, prompting a scholarly debate that revolves around the impacts of this increasingly popular tourism form.

Niche Tourism Impacts

The sustainability argument underlying niche tourism is based on the potential socio-economic and environmental benefits it may offer to destinations. Niche tourism has been argued to provide significant opportunities for tourism development in small remote and peripheral destinations that often lack resources (Butler, 2019). In this capacity, niche tourism may lead to a sustainable form of tourism development that will benefit such destinations in the long term. Peripheral areas, in particular, may diversity their income sources through niche tourism which is often regarded as a panacea to the economic decline and youth migration experienced in such destinations (Dinis and Krakover, 2016). Niche tourism also enables established destinations to diversify their tourism product offering by introducing year-round niche tourism forms and therefore addressing seasonality (Malik, 2018; Garanti, 2022). Moreover, given the small-scale nature of niche tourism, it may offer greater control to the local community over tourism development. In turn, locals may engage more actively in tourism planning and utilize local products, networks and knowledge in an effective manner, thereby enhancing the destination's sustainability potential.

In addition to encouraging local community participation in tourism, niche tourism may lead to an improvement in locals' well-being. As niche tourism is based on smaller-scale development, it may help to mitigate the negative impacts of over-commercialization and overcrowding on destinations (Boukas and Chourides, 2016). Likewise, due to its smaller-scale operation and emphasis on environment, niche tourism may promote the conservation of the environment (Wolf *et al.*, 2017) and minimize the harmful effects that tourism development may yield on the ecosystem. Last, niche tourism may offer benefits to tourists as well through the provision of greater satisfaction and deeper engagement with the destination's environment and culture (Swanson and Cavender, 2019), making the tourist experience more authentic, meaningful and memorable (Sharma and Nayak, 2020). On the other hand, niche tourism has been criticized for its emphasis on specialization and small targeted market segments which may limit its economic contribution to destinations, especially when considering the relatively high cost of development of certain niche tourism products (Farmaki, 2015). Therefore, the economic benefits for local communities are questionable, particularly if they lack the necessary skills and experience to cater to niche tourists, suggesting that niche tourism may not necessarily provide a more economically sustainable alternative to mass tourism for destinations. Critics also raised questions on niche tourism's ability to improve the livelihood of the local community. Indeed, niche tourism tends to commodify the local culture as well as natural resources as a large proportion of niche tourism products are based on social and environmental resources. Correspondingly, an overuse of natural resources may take place that will in turn put a strain on the environment (Butler, 2019; Khanra *et al.*, 2021) while the local community may experience a cultural degradation as elements of the local culture transform into a competitive culture tourism product (Coronado, 2014; Zou *et al.*, 2021).

Also, the popularity of niche tourism entails that its offering is not confined to small independent companies as many large tour-operating companies are designing and promoting niche tourism products to larger market segments (Wilhelm Stanis and Barbieri, 2013). In relation to this point, it may be said that the vast range of products offered under the label niche tourism require deep understanding of each niche tourism market if the long-term viability of the product is to be maintained. Yet, failure to do so by larger tour operators implies that niche tourist behaviours are not well considered in the development of niche tourism, leading to unstable demand and poor satisfaction rates. The provision of niche tourism by large tour operators also suggests limited economic benefits to the local community and minimal local participation in tourism planning. Evidently, the boundary between mass tourism and niche tourism is rather blurred (Agarwal *et al.*, 2018). Indeed, Ma *et al.* (2020) argued that a large share of tourists involved in what is traditionally understood as niche tourism activities exhibit behaviour and profile characteristics of mass tourists seeking novelty.

These challenges pose the question whether niche tourism may in fact be considered a contributor to destinations' sustainability, highlighting the need for further research on the niche tourism and sustainability relationship. Novelli *et al.* (2022) argued that the impacts of niche tourism compared to mass tourism from a sustainability perspective remain an under-explored research area. Indeed, most studies have looked at various niche tourism forms from a competitiveness angle, exploring tourist satisfaction and behaviour (e.g. Sharma and Nayak, 2020; Cillo *et al.*, 2021), or a developmental aspect delving into strategies and practices (Koens *et al.*, 2009; Nair and Mohanty, 2021). In addition, the focus of past studies on specific niche tourism forms has led to a rather fragmented literature which adds to the complexity of the examination of the niche tourism and sustainability nexus. More research is, thus, required on how specific niche tourism forms may contribute to sustainability and, in particular, to the Sustainable Development Goals outlined in the United Nations' Agenda 2030 amidst global challenges such as changing consumer needs post COVID-19 and technological advancements. This edited volume hopes to fill in this gap in knowledge and enhance understanding of niche tourism and sustainability aspects.

Structure of the Book

The book aims to contribute to the debate on niche tourism and sustainability. Specifically, through a collection of 12 conceptual and empirical chapters presenting niche tourism cases from around the world, the book seeks to add to the discussion on the niche tourism and sustainability interface. The book is structured around the following themes: (i) the emerging trends shaping demand characteristics and behaviours; (ii) resident perspectives of niche tourism; (iii) niche tourism implementation practices; and (iv) prospects of niche tourism in terms of sustainability. Therefore, the book offers a comprehensive overview of the dynamic relationship between niche tourism and sustainability. The book also serves as a platform for critical inquiry of aspects pertinent to niche tourism and sustainability and, in turn, imparts important theoretical and practical insights to destination planners, policy makers, industry practitioners, educators and students on how niche tourism can become more sustainable and also assistive in strengthening destinations' sustainability goals.

Part I consists of three chapters discussing tourist behavioural aspects and experiences that shape demand for niche tourism. Chapter 1, authored by Beverly Ntshabeleng and Madiseng Phori, explores the transition from mass tourism to sustainable niche tourism in South Africa, highlighting the emerging importance of mindful tourism in sustaining destinations. By examining South Africa, the chapter discusses how integrating ecotourism, cultural preservation, community engagement and mindfulness can enhance tourist satisfaction and improve sustainability. In Chapter 2, Feras Orekat, Glenn Murray and Andrew Le underline the significance of visitor experiences in heritage tourism development. By exploring visitor encounters at the heritage site of Sydney's Quarantine Station, the chapter identifies cultural characteristics as significantly influential in shaping visitors' overall experiences. Werner Gronau, Jana Große Hokamp and Knut Scherhag authored Chapter 3 which addresses the preconditions of motorcycle tourism in regard to the three pillars of sustainability. The chapter analyses the attitudes and motivations of motorcycle tourists and outlines the developmental requirements for destinations in making motorcycle tourism more sustainable.

In Part II, which examines residents' perspectives of niche tourism, two chapters are presented. Chapter 4 is authored by Fani Efthymiadou and Chryso Panayidou and it discusses resident perceptions and support for zero waste management as part of environmental practices in the context of ecotourism. The study draws from the island of Tilos (Greece) to reveal that residents demonstrate a high level of environmental awareness and support towards a zero waste programme implemented on the island. In this context, residents' perceptions and attitudes enhance the destination's ecotourism image as a sustainable niche tourism form. Emrullah Erul and Furkan Atasoy Karacabey authored Chapter 5, which investigates how residents' attachment to their community and perceptions of mass tourism influence their attitudes towards niche tourism development. Using the stimulus–organism–response theory, the study draws from Manavgat (Türkiye) to examine how resident attitudes predict their support for niche tourism in the light of expectations of potential benefits.

Part III includes four chapters that discuss niche tourism implementation practices amidst the challenges destinations face in terms of sustainability. In Chapter 6, Athina Nella and Dimitrios Stergiou analyse UK stakeholder practices in terms of birdwatching tourism and reveal varied sustainability efforts, ranging from eco-friendly travel options to community engagement initiatives. Chapter 7, authored by Stella Kladou, Maria Alebaki and Kleio Arvanakou, draws from the island of Corfu in Greece to explore sustainability practices and challenges for family wineries in the context of wine tourism. In Chapter 8, Sıla Karacaoğlu examines the role of creative tourism in the sustainable development of rural areas. By presenting cases of creative tourism practices in Türkiye, the author offers a comprehensive overview of creative tourism as an important niche tourism form that enhances and sustains the tourist appeal of rural destinations. Maria Hadjielia Drotarova and Vasilis Papavasiliou are the authors of Chapter 9, which explores the practices of rural small and medium tourism enterprises (SMTEs) in Cyprus. The chapter considers digital transformation as an innovative way to deal with the challenges encountered by rural SMTEs, thus promoting sustainable rural tourism development.

In Part IV, the prospects of niche tourism in terms of contributing to the sustainable development of destinations are considered through the presentation of three chapters. In Chapter 10, through a discussion of the Israel/Palestine conflict, Katerina Antoniou examines the role of citizen diplomacy enacted by residents and tourists in fostering peace tourism, which may contribute to the destinations' sustainable development. Gülsel Çiftci authored Chapter 11 in which she discusses the impact of cross-border tourism on social cohesion, justice and peace, advancing understanding of how cross-border tourism can contribute to sustainable development of destinations and the maintenance of peaceful societies. Chapter 12, authored by Ângela Gonçalves, Jéssica Ribeiro, Mariana Pereira and Bruno Barbosa de Sousa, is the last chapter of this book and offers an examination of volunteer tourism as an emerging sustainable niche tourism form that may exert positive impacts on local communities and support destinations' sustainability goals.

Undoubtedly, the list of topics presented and discussed in this edited volume is not exhaustive of the important theme of niche tourism, which includes a huge range of tourism products that is likely to further expand as destinations strive for uniqueness and innovativeness as a competitive advantage. Even so, through a collection of selective cases, this book introduces the possibility of niche tourism in contributing to the sustainability potential of destinations. As such, the book may offer important theoretical insights to researchers aspiring to delve deeper into the examination of the niche tourism and sustainability nexus. Practical recommendations may also be provided to destination stakeholders such as tourism planners and industry practitioners wishing to align policies and practices to sustainability principles through the development of niche tourism.

References

Agarwal, S., Busby, G. and Huang, R. (eds) (2018) Special tourism: an introduction. In: *Special Interest Tourism: Concepts, Contexts and Cases*. CAB International, Wallingford, UK, pp. 1–17.

Boukas, N. and Chourides, P. (2016) Niche tourism in Cyprus: conceptualising the importance of social entrepreneurship for the sustainable development of islands. *International Journal of Leisure and Tourism Marketing* 5(1), 26.

Bunghez, C.L. (2021) The emerging trend of niche tourism: impact analysis. *Journal of Marketing Research and Case Studies* 2021, 134710.

Butler, R.W. (2019) Niche tourism (birdwatching) and its impacts on the well-being of a remote island and its residents. *International Journal of Tourism Anthropology* 7(1), 5.
Cillo, V., Rialti, R., Del Giudice, M. and Usai, A. (2021) Niche tourism destinations' online reputation management and competitiveness in big data era: evidence from three Italian cases. *Current Issues in Tourism* 24(2), 177–191.
Coronado, G. (2014) Selling culture? Between commoditisation and cultural control in indigenous alternative tourism. *PASOS Revista de Turismo y Patrimonio Cultural* 12(1), 11–28.
Dinis, A. and Krakover, S. (2016) Niche tourism in small peripheral towns: the case of Jewish heritage in Belmonte, Portugal. *Tourism Planning & Development* 13(3), 310–332.
Farmaki, A. (2012a) A supply-side evaluation of coastal tourism diversification: the case of Cyprus. *Tourism Planning & Development* 9(2), 183–203.
Farmaki, A. (2012b) An exploration of tourist motivation in rural settings: the case of Troodos, Cyprus. *Tourism Management Perspectives* 2–3, 72–78.
Farmaki, A. (2015) Regional network governance and sustainable tourism. *Tourism Geographies* 17(3), 385–407.
Garanti, Z. (2022) Alternative and special interest tourism to mitigate the effects of tourism seasonality: the debate from Cyprus. *Worldwide Hospitality and Tourism Themes* 14(5), 451–460.
Khanra, S., Dhir, A., Kaur, P. and Mäntymäki, M. (2021) Bibliometric analysis and literature review of ecotourism: toward sustainable development. *Tourism Management Perspectives* 37, 100777.
Koens, J.F., Dieperink, C. and Miranda, M. (2009) Ecotourism as a development strategy: experiences from Costa Rica. *Environment, Development and Sustainability* 11(6), 1225–1237.
Lew, A.A. (2008) Long tail tourism: new geographies for marketing niche tourism products. *Journal of Travel & Tourism Marketing* 25(3–4), 409–419.
Ma, S. (David), Kirilenko, A.P. and Stepchenkova, S. (2020) Special interest tourism is not so special after all: big data evidence from the 2017 Great American Solar Eclipse. *Tourism Management* 77, 104021.
Malik, G. (2018) Niche tourism: a solution to seasonality. *Journal of Management Research and Analysis* 5(2), 80–93.
Marson, D. (2011) From mass tourism to niche tourism. In: Robinson, P., Heitmann, S. and Dieke, P. (eds) *Research Themes for Tourism*. CAB International, Wallingford, UK, pp. 1–15.
McCabe, S. (2015) Are we all post-tourists now? Tourist categories, identities and post-modernity. In: Singh, T.V. (ed.) *Challenges in Tourism Research*. Channel View Publications, Bristol, UK, pp. 18–26.
Nair, B.B. and Mohanty, P.P. (2021) Positioning spice tourism as an emerging form of special interest tourism: perspectives and strategies. *Journal of Ethnic Foods* 8(1), 10.
Novelli, M. (2005) *Niche Tourism: Contemporary Issues, Trends and Cases*. Elsevier, Oxford.
Novelli, M. (2018) Niche tourism: past, present and future. In: Cooper, C., Volo, S., Gartner, W.C. and Scott, N. (eds) *The SAGE Handbook of Tourism Management*. SAGE Publications Ltd, London, pp. 344–359.
Novelli, M., Cheer, J.M., Dolezal, C., Jones, A. and Milano, C. (eds) (2022) Introduction to niche tourism – contemporary trends and development. In: *Handbook of Niche Tourism*. Edward Elgar Publishing, Cheltenham, UK, pp. xxiii–xxxii.
Papathanassis, A. (ed.) (2011) Revisiting the tourism long-tail scenario: epilogue. In: *The Long Tail of Tourism: Holiday Niches and Their Impact on Mainstream Tourism*. Gabler Verlag/Springer Fachmedien, Wiesbaden, Germany, pp. 213–220.
Richards, G. (2021) Rethinking niche tourism: the example of backpacking. *Croatian Regional Development Journal* 2(1), 1–10.
Sharma, P. and Nayak, J.K. (2020) Examining experience quality as the determinant of tourist behavior in niche tourism: an analytical approach. *Journal of Heritage Tourism* 15(1), 76–92.
Skinner, H.M. and Soomers, P. (2019) Spiritual tourism on the island of Corfu: positive impacts of niche tourism versus the challenges of contested space. *International Journal of Tourism Anthropology* 7(1), 21.
Swanson, J.R. and Cavender, R. (2019) Generational perceptions of prosperity on the niche tourism island destination of Ikaria, Greece. *International Journal of Tourism Anthropology* 7(1), 40.
Trunfio, M., Petruzzellis, L. and Nigro, C. (2006) Tour operators and alternative tourism in Italy: exploiting niche markets to increase international competitiveness. *International Journal of Contemporary Hospitality Management* 18(5), 426–438.
Wilhelm Stanis, S.A. and Barbieri, C. (2013) Niche tourism attributes scale: a case of storm chasing. *Current Issues in Tourism* 16(5), 495–500.

Wolf, I.D., Ainsworth, G.B. and Crowley, J. (2017) Transformative travel as a sustainable market niche for protected areas: a new development, marketing and conservation model. *Journal of Sustainable Tourism* 25(11), 1650–1673.

Zou, Y., Meng, F., Bi, J. and Zhang, Q. (2021) Evaluating sustainability of cultural festival tourism: from the perspective of ecological niche. *Journal of Hospitality and Tourism Management* 48, 191–199.

Part I: Tourist Behaviour and Experiences

This first part of the edited book consists of three chapters discussing tourist behavioural aspects and experiences. The first chapter (Chapter 1), authored by Beverly Ntshabeleng and Madiseng Phori, explores the transition from mass tourism to sustainable niche tourism in South Africa, highlighting the importance of a mindful tourist behaviour in promoting destination sustainability. The second chapter (Chapter 2) by Feras Orekat, Glenn Murray and Andrew Le draws from Sydney's Quarantine Station to demonstrate the significance of visitor experiences in developing heritage tourism. Werner Gronau, Jana Große Hokamp and Knut Scherhag authored the third and final chapter of this part (Chapter 3); it analyses the attitudes and motivations of motorcycle tourists in order to provide greater understanding of how tourism demand aspects can be incorporated in the effort to develop a more sustainable form of motorcycle tourism.

1 Exploring Niche Tourist Behaviour in the Global South: Dynamics and Prospects in South Africa

Beverly M. Ntshabeleng* and Madiseng M. Phori
Department of Tourism Management, Faculty of Management Sciences, Tshwane University of Technology, Pretoria, South Africa

Abstract
This chapter explores tourism's transition from mass tourism to sustainable niche tourism, aligning with post COVID-19 shifts towards mindful travel in the Global South, with a specific focus on South Africa. It highlights a growing demand for specialized experiences and emphasizes the importance of mindful tourism in sustaining destinations. Despite the prevailing focus on mass tourism, there is a call for a broader perspective on niche tourism to benefit both tourists and local communities economically and environmentally. The study advocates for tailored offerings that reflect tourists' evolving preferences, promoting sustainability and positive contributions to destinations. By linking niche tourism strategies to sustainability, the chapter underscores how sustainable practices can enhance the economic and environmental resilience of tourist destinations. Additionally, it discusses the implications of this transition for other contexts beyond South Africa, suggesting that the principles of sustainable niche tourism can be applied to various regions globally. This approach aims to position South African tourism as sustainable, fulfilling and culturally rich, meeting the expectations of mindful tourists while preserving the integrity of destinations.

Keywords: Niche tourism, sustainable tourism, mindful tourist, South Africa, Global South

Introduction

Niche tourism markets in the Global South present both challenges and prospects in terms of the dynamics of tourist behaviour. Understanding niche markets as social rituals, involving consumers and producers, is crucial for sustainable development in tourism (Richards, 2021). This chapter investigates an essential aspect of contemporary tourism, the transition from mass tourism to sustainable and mindful niche tourism. This shift is imperative in meeting the evolving demands of experienced tourists, particularly in the post COVID-19 era, characterized by heightened awareness and preferences for specialized vacation experiences.

South African tourism can transition from a mass tourism focus to a more sustainable and mindful niche tourism approach by incorporating ecotourism projects in urban areas, developing tourism routes in local municipalities and addressing the impact of climate change on tourism demand. Urban ecotourism destinations, such as Dinokeng in the Tshwane metropolitan area, offer a unique opportunity

*Corresponding author: NtshabelengBM@tut.ac.za

DOI: 10.1079/9781800626669.0001

to combine wildlife conservation with urban development (Burton and Rogerson, 2023). The development of tourism routes, as seen in the City of Mbombela Local Municipality, can contribute to community development and economic growth while preserving the sociocultural and environmental attributes of the area (Sekole *et al.*, 2022).

The central question guiding this exploration is how tourism industries in the Global South, especially in countries such as South Africa, can effectively align their offerings with the dynamics of tourist behaviour, emphasizing mindfulness and sustainability. Chong (2020) indicates that there is a noticeable shift in tourist preferences towards niche experiences, signalling a departure from traditional mass tourism models. This trend underscores the need for destination management strategies that cater to the growing demand for mindfulness-oriented tourism (Richards, 2021). Therefore, the purpose of this chapter is to understand the niche tourism dynamics and strategies that foster mindful tourism experiences. Furthermore, this chapter contributes to the ongoing discourse on sustainable tourism by highlighting the importance of niche tourism in meeting the changing expectations of modern tourists while ensuring the preservation and enhancement of cultural and natural heritage. The chapter offers valuable insights and recommendations for stakeholders in the tourism industry to navigate the dynamics and opportunities presented by the evolving tourism landscape in the Global South. By understanding and responding to these evolving demands, destinations can not only attract tourists but also ensure the long-term sustainability of their cultural and environmental assets.

Theoretical Background

There has been a pronounced shift in tourist behaviour with an increasing demand for specialized vacation experiences among seasoned tourists (Kachniewska, 2014; Di-Clemente *et al.*, 2020). Tourists are now seeking emotional and experiential products, leading tourism organizations to employ fresh strategies to meet these new demands (Mihajlović, 2020). This shift is driven by factors such as socio-economic changes, specific tourist motives (i.e. adventures, new experiences and culture) and economic changes reflected in purchasing power (Chhabra, 2022). Additionally, the invention of Web 2.0 and the rise of social media have changed the way tourists extract information and perceive their travel experiences, emphasizing the importance of various social media platforms for both tourists and the tourism industry (Somrak and Khaewmesuan, 2022).

Moreover, the behaviour of tourists and their preferences for sustainable niche tourism can be shaped by various significant factors. Firstly, Lu and Nepal (2009) found that the availability of reliable information on the sustainability of destinations positively influences the adoption of pro-sustainable travel and tourism habits. Secondly, as highlighted by Kollmuss and Agyeman (2002), personal factors such as environmental values, attitudes and habits shape tourists' environmentally friendly behaviour. Additionally, Poria *et al.* (2004) discussed how destination context including facilitating conditions, tourist identity and culture, and visual cues influences tourist behaviour by conveying social norms related to environmental friendliness. Ajzen and Fishbein (1980) emphasized that psychological determinants such as personality characteristics, attitudes, evaluations and emotions influence tourist behaviour and demand. Furthermore, Steg and Vlek (2009) noted that factors like environmental knowledge, participation motivation, opportunity, self-identity and commitment have significant positive effects on tourists' pro-environmental behaviour. These factors collectively shape tourist behaviour and their demands for sustainable niche tourism. The COVID-19 pandemic has further accelerated these changes, with online information sources taking precedence over consulting friends and relatives, and a significant advance in digitization is expected, leading to the displacement of physical travel agencies by online platforms.

A critical aspect of this chapter is the synthetic analysis of existing literature on niche tourism and sustainability. By synthesizing and critically evaluating previous studies, the chapter aims to provide insights into the dynamics and opportunities facing the tourism industry in transitioning towards a more niche-oriented and sustainable paradigm. The methodology employed involves a thorough synthesis analysis of scholarly works, industry reports and case studies to extract valuable lessons and best practices. The implications of this research are significant for destination management, policy formulation and tourism development strategies, particularly in areas grappling with the tension between mass tourism-driven economic growth and the need for sustainable practices. Adopting a sustainable tourism approach that considers the 'Triple Bottom Line' framework (a sustainability framework proposing that businesses should consider social and environmental issues along with profit) and addresses carbon emissions can ensure long-term sustainable growth for the South African tourism industry (Proos and Hattingh, 2022).

Methodology

The research methodology for the present study includes a conceptual framework about sustainable niche tourism, synthesizing previous studies and industry reports (Kumar *et al.*, 2018). The significance of a conceptual paper is its ability to develop logical and complete arguments rather than to test them empirically. In this chapter, a research design rooted in conceptual theory synthesis was employed to enhance understanding of the specific concept studied, which is sustainable niche tourism. Theory synthesis refers to the conceptual integration across multiple theoretical perspectives (Jaakkola, 2020). The potential goal of this approach is to summarize and integrate current understanding on a topic, thereby advancing knowledge on the interface between niche tourism and sustainability.

Specifically, the present study involved a search for studies on sustainable niche tourism and mindful tourist experiences across databases such as ResearchGate, Wiley Online Library Full Collection, Science Direct, Taylor & Francis Online, Scopus (A&I), Springer LINK and Google Scholar. This approach contributes to the ongoing discourse on sustainable tourism by highlighting the significance that niche tourism plays in meeting modern tourists' expectations while preserving and enhancing cultural and natural heritage (de Bruyn *et al.*, 2023). A critical discussion of the literature follows, building around the main themes that emerge from the review of pertinent studies and which relate to the development of sustainable niche tourism in the context studied. In particular, tourist behaviour in relation to sustainable niche tourism is discussed in addition to the importance of targeting mindful tourists in shaping a sustainable tourist experience.

Sustainable Niche Tourism in South Africa

South Africa's tourism industry is enriched by a diverse array of sustainable niche tourism types, each contributing uniquely to the country's cultural, environmental and economic landscapes. From ecotourism to culinary experiences, these niche tourism segments showcase South Africa's rich heritage while promoting sustainability and community engagement.

Sustainable niche tourism in South Africa plays a crucial role in the country's tourism landscape. Ecotourism stands out for its focus on environmental conservation and biodiversity protection, engaging tourists in activities like guided nature walks and wildlife viewing while emphasizing the importance of ecological sustainability (Nel and Rossouw, 2020). Cultural heritage tourism, on the other hand, immerses tourists in the country's rich cultural heritage through traditional dances, craft demonstrations and visits to cultural sites, fostering cultural exchange and supporting local artisans (Rogerson, 2020). For adventure seekers, South Africa offers a plethora of activities under adventure tourism including hiking, mountain biking and water sports, all set in diverse natural landscapes that prioritize safety and environmental conservation (Van Niekerk, 2022). Township tourism emerged post the 1994 democratic elections in South Africa, spurred by the recognition of townships as politically significant areas appealing to tourists interested in the country's transition to democracy (Ramchander, 2007; Vos, 2015). These townships are scattered across all major cities in each of the nine provinces (Ramchander, 2007), with Johannesburg and Cape Town particularly popular for township tours, representing a profitable venture for tour operators (Vos, 2015).

Additionally, wellness tourism caters to tourists seeking relaxation and holistic well-being, with spa retreats, yoga sessions and wellness workshops set against the country's scenic beauty and serene environments (United Nations, 2018). Furthermore, agritourism and food tourism highlight South Africa's agricultural prowess and culinary traditions, offering farm visits, agricultural activities and culinary experiences that promote sustainable farming practices and celebrate local flavours (Van Niekerk, 2022). These niche tourism types collectively contribute to environmental conservation, cultural preservation and community development while enhancing the overall tourism experience in South Africa. Sustainable niche tourism in South Africa is not just about offering unique experiences to tourists but also about preserving natural habitats, safeguarding cultural heritage, supporting local economies and promoting sustainable practices. By embracing and further developing these niche tourism segments, South Africa can continue to thrive as a leading destination for responsible and authentic tourism experiences, contributing to the country's long-term sustainability and growth in the tourism sector.

Sustainable niche tourism in South Africa has garnered increasing scholarly attention in recent years. Scholars such as Rogerson and Rogerson (2021a, b) emphasize the importance of sustainable practices in niche tourism, particularly in South Africa's diverse ecological and cultural landscapes. The authors argue that sustainable niche tourism not only contributes to economic development but also fosters environmental conservation and community empowerment (Rogerson and Rogerson, 2021a, b). Their research highlights how niche tourism initiatives can create job opportunities, promote local entrepreneurship and encourage the conservation of natural and cultural resources. Building on this perspective, Ntshingila and Musakwa (2020) explore the role of community-based tourism enterprises (CBTEs) in promoting sustainable niche tourism. Their study highlights the potential of CBTEs to empower local communities economically while preserving cultural heritage and natural resources. Key findings indicate that CBTEs can serve as a model for sustainable tourism development by integrating local community participation and ensuring that tourism benefits are equitably distributed. Similarly, Singh (2021) focused into the significance of cultural heritage in sustainable niche tourism, emphasizing the need for responsible tourism practices that respect and promote local cultures. The study also suggests that cultural heritage tourism can drive economic growth by attracting niche market tourists who are willing to pay a premium for unique and meaningful experiences. By providing these various niche tourism options, South Africa's tourism industry seeks to meet different tourist demands and encourage tourists to take actions that will contribute towards sustainable niche tourism.

Tourist demand and behaviour in sustainable niche tourism

The primary focus of niche tourism is on the demands of the tourists (Bunghez, 2021). Tourists are consistently seeking a more specialized and fulfilling tourism experience. This necessitates tourism organizations in South Africa to create inventive products to stay competitive. According to Bunghez (2021), when considering tourist demand, tourism organizations in the Global South – particularly South Africa – must actively engage tourists in a specific niche through their unique consumerist strategies. These strategies must contribute to ensuring that critical factors such as tourism facilities (e.g. activities, attractions, settlements and food), hospitability, ambience, surrounding environments and crowd density meet tourist demand (Mutum and Ghazali, 2023).

While sustainability is typically not the foremost consideration when choosing to engage in tourism, there is a growing trend among tourists to prioritize sustainable tourism. Approximately one-third of tourists are interested in sustainable tourism or specific aspects of sustainability (Weber, 2019). Poon observed this shift as early as 1989, when tourists began to demand more sustainable yet segmented, customized and flexible vacations rather than standardized holiday packages (Poon, 1989). In the same notion, Novelli (2018) concurs that there is a growing demand for conscious and transformative vacations, a trend in which sustainable tourism can provide the means to change both lifestyles and increase the positive caring impact made on destinations.

In South Africa, the increasing demand for sustainable tourism may be attributed to the advancements in modern systems of representation, transportation, communication and digitization (Mayer *et al.*, 2023). Moreover, for the first 2 years of the COVID-19 pandemic, the majority of individuals globally including South Africa were confined to their homes, placed in lockdowns and socially isolated. This has prompted an interest in engaging with others in more expansive, 'greener' environments (Mell and Whitten, 2021). The pandemic has changed tourist demands, presenting tourism organizations with an opportunity to shift away from mass tourism and develop innovative niche tourism offerings (Azmi *et al.*, 2023).

According to Sert (2017), niche tourism offerings should be distinctive and not have many alternatives for individual tourists; they are tailored to a group of tourists with similar characteristics. In the mid-1990s, Dalgıç and Leeuw (1994) cautioned that when assessing the viability of a niche market, service providers should consider the following factors:

- The niche market should have a size that is sufficient to potentially generate profits.
- New entrants should have a competitive advantage.
- It is imperative not to overlook potential competitors.
- There should be growth potential.
- There should be desire for special treatment.
- Customer goodwill.
- There should be a market that has enough purchasing power.

It is necessary to evaluate these factors in context with regions, as these factors may affect behaviours of niche tourists from developed regions in a distinct manner compared to those from developing or underdeveloped regions (i.e. the Global South). For instance, empirical research indicates that niche tourists from the developed countries have higher purchasing power (Lissner and Mayer, 2020) compared to those from the Global South regions (Gisore and Ogutu, 2015; Mgonja *et al.*, 2017). Accordingly, the Global South regions should assess these factors so as to figure out how to enhance niche tourism's accessibility, not only for international tourists but also for local tourists. Moreover, South Africa needs to address knowledge gaps regarding the changing supply–demand situations of specific niche tourism offerings post COVID-19 to inform evidenced-based policy interventions, decision making and strategy formulation (Rogerson and Rogerson, 2021a, b). Given the specialized nature of niche tourism, it is critical for niche tourism interventions to comprehend what can add value and enhance tourist experiences in order to foster mindfulness.

Mindful tourists

Mindfulness can be defined as 'a state of conscious awareness in which the individual is implicitly aware of the context and content of information' (Langer, 1992, p. 289). Mindfulness theory delineates two distinct modes of information processing, namely mindful and mindless (Langer, 1989). A mindful approach is characterized by its openness and capacity for elaborate thinking, while a mindless approach relies on inflexibility and less elaborate thinking, often based on past distinctions (Luttrell *et al.*, 2014). In contrast to mindfulness, Moscardo (2017) views mindlessness as a single-minded reliance on information without an active awareness of alternative perspectives or uses for the information received. Mindfulness enables individuals to perceive situations with clarity and recognize underlying possibilities, rather than being confined by preconceived notions (Shapiro, 2020). Dutt and Ninov (2016) posit that promoting mindfulness among tourists in the Global South, including South Africa, would be advantageous due to its association with numerous beneficial outcomes.

Mindfulness in tourism has been linked to interpretation, demonstrating the efficacy and significance of interpretation in tourists' experiences and acquired knowledge (Moscardo and Pearce, 1986; Eck *et al.*, 2023). This approach explores how tourists build interpretations to enhance the overall quality of their experience and raise awareness of their behaviour (Suryandari and Rahmawati, 2022). Benefits of mindfulness for tourists include enhanced learning and comprehension, more enjoyable and positive tourism experiences (Lee and Kim, 2018; Taylor and Norman, 2019), improved recollection of information (Moscardo and Pearce, 1986; Dutt and Ninov, 2016) and psychological and physical well-being (Chen *et al.*, 2017; Wang *et al.*, 2021).

In South Africa and other parts of the Global South, scholars suggest that mindfulness could benefit tourism destinations by increasing tourists' awareness and concern, thereby aiding in destination sustainability (Moscardo, 1996; Stankov and Filimonau, 2023). Given this background, it is reasonable to deduce that the practice of mindfulness in tourism contributes to achieving Goal 12 of the United Nations' Sustainable Development Goals 2030, which focuses on ensuring sustainable consumption and production patterns (United Nations, 2015). Tourism organizations in the Global South can intentionally craft tourist experiences to induce mindfulness or facilitate profound shifts in consciousness, both during and after the visit (Sheldon, 2020).

Sustainable tourist experience

The concept of experience has garnered significant interest from various academic fields. In psychology, experience is a fundamental aspect of consciousness that is consistently present throughout a person's life (Gadamer, 1979). Drawing on this definition, tourist experience is a subjective mental state (O'Dell and Billing, 2005; de Freitas Coelho *et al.*, 2018) and multisensory performance, relying on tourists' specific interests, motivations, perceptions, depictions, level of engagement and sharing while participating in tourism activities (Stankov *et al.*, 2022). Experience includes learning, enjoyment and escapism (Kang and Gretzel, 2012). The 'learning experience' refers to the process by which a tourist acquires new knowledge and skills (Pearce, 2005; Câmara *et al.*, 2022). The 'enjoyment experience' refers to the degree to which the tourist experience is deemed enjoyable beyond its practical value (Davis *et al.*, 1992). The 'escape experience' refers to the immersive encounter that tourists have with the destination environment, allowing them to detach from their everyday lives (Pearce, 2005).

The level of learning, enjoyment and escapism is determined not only by the tourist but also by tourism organizations (Moreira *et al.*, 2021), converging tourism consumption and production to create value (Akel, 2022). Within this context, Stankov *et al.* (2022) suggest that implementing mindfulness is an effective approach for tourism organizations to integrate individual tourist demands into a value proposition, ultimately enhancing the meaningfulness of tourists' experiences. Rubin *et al.* (2016) highlight that in addition to recognizing the unique demands of tourists, tourism organizations should acknowledge the factors that enable tourists to be engaged and have meaningful experiences.

The framework proposed by Walls *et al.* (2011) identifies four key factors that influence the overall experience of consumers (tourists): (i) physical environment, which refers to the physical facilities and surroundings; (ii) human interaction environment, referring to the interactions that occur between tourists and tourism organizations; (iii) situational factors, which refers to specific attributes that shape the characteristics of the visit, such as the purpose of the visit and the individuals accompanying the tourist; and (iv) individual characteristics, including personality traits and sensitivity to surroundings, influencing how tourists perceive the coordinated tourism services. Collectively, these factors contribute to the overall tourist experience (Zhang *et al.*, 2016). If one factor falls short, others can compensate, thus enhancing mindfulness in the tourist experience. The effective implementation of mindfulness into tourist experiences requires collaborative efforts between tourists and tourism organizations to integrate these factors meaningfully.

In the context of the Global South, involving the local community in developing niche tourism is crucial for creating a mindful experience. Kuntai (2020) suggests that in Kenya's Maasai Mara National Reserve, local community involvement is essential. Moscardo's (2017) study in South Africa's Kruger National Park and Soweto revealed that storytelling plays a significant role in encouraging mindful tourists and supporting positive tourist experiences. Manwa *et al.*'s (2017) study indicated that when tourism organizations fail to offer experiences fostering mindfulness, it may be difficult for niche tourism, such as cultural tourism, to grow into a significant economic driver in the Global South (e.g. Lesotho). Sarantou *et al.* (2021) discovered that in Namibia, niche tourism offerings promoting mindfulness not only develop sustainable livelihoods for local service providers but also offer highly meaningful experiences for tourists.

Mindfulness in tourism experiences benefits tourists by meeting their needs for enhanced experiences, potentially leading to increased tourist satisfaction (Eck *et al.*, 2023) and raising awareness of their behaviour (Suryandari and Rahmawati, 2022). This, in turn, helps tourism providers achieve their business goals (e.g. profitability) (Suryandari and Rahmawati, 2022) while assisting local communities in preserving their natural and cultural resources for present and future generations and gaining economic benefits (e.g. employment) (Kuntai, 2020).

Conclusions and Implications

The purpose of the present study was to explore the transition from mass tourism to sustainable niche tourism, emphasizing the importance of creating meaningful and mindful tourist experiences. The study focused on South Africa due to its rich cultural and natural diversity, which provided an ideal context for developing sustainable niche tourism. By examining South Africa, the study aimed to illustrate how integrating ecotourism, cultural preservation, community engagement and mindfulness can enhance tourist satisfaction, support local economies and contribute to long-term environmental and social benefits. Insights from South Africa's approach can offer valuable lessons for other regions in the Global South aiming to adopt sustainable tourism practices.

The growing demand for specialized vacation experiences among seasoned tourists post COVID-19 underscores the need for destinations to align with evolving tourist behaviour and preferences (Kachniewska, 2014; Di-Clemente *et al.*, 2020). To achieve this transition successfully, several key recommendations emerge and are presented below.

1. *Incorporating ecotourism projects*: South African tourism can incorporate ecotourism projects in urban areas to create unique experiences that combine wildlife conservation with urban development. This approach attracts mindful tourists and contributes to environmental conservation efforts (Burton and Rogerson, 2023).
2. *Developing tourism routes*: The development of tourism routes in local municipalities, such as those seen in the City of Mbombela Municipality, can promote community development, economic growth and cultural preservation. These routes offer tourists immersive experiences while supporting local economies (Sekole *et al.*, 2022).
3. *Addressing climate change*: Addressing the impact of climate change on tourism demand is crucial for sustainable tourism growth. By adopting sustainable tourism practices and considering the 'Triple Bottom Line' framework, South Africa can mitigate negative environmental impacts while ensuring long-term growth (Proos and Hattingh, 2022).
4. *Responding to evolving tourist demands*: Understanding and responding to evolving tourist demands, including a preference for emotional and experiential products, is essential for tourism operators. This entails developing fresh strategies, leveraging advancements in communication technologies and embracing sustainable practices (Mihajlović, 2020; Somrak and Khaewmesuan, 2022).
5. *Promoting mindfulness*: Promoting mindfulness among tourists can enhance their experiences and contribute to sustainable tourism practices. By creating meaningful and immersive experiences that foster mindfulness, tourism organizations can increase tourist satisfaction and promote destination sustainability (Moscardo and Pearce, 1986; Sheldon, 2020; Suryandari and Rahmawati, 2022).

In this chapter, we focused on South Africa to illustrate how sustainable niche tourism can be developed in a context rich with cultural and natural diversity. The South African example highlights the potential for integrating ecotourism, cultural preservation, community engagement and mindfulness into a cohesive strategy for sustainable tourism. However, the insights from the case study can be applied more broadly to the Global South and beyond. This chapter demonstrates that by adopting these practices, destinations can create sustainable and meaningful tourist experiences that contribute to long-term environmental, economic and social benefits.

References

Ajzen, I. and Fishbein, M. (1980) *Understanding Attitudes and Predicting Social Behavior*. Prentice-Hall, Englewood Cliffs, New Jersey.

Akel, G. (2022) Destination's image and tourism experiences. In: da Costa, R.A., Brandão, F., Breda, Z. and Costa, C. (eds) *Planning and Managing the Experience Economy in Tourism*. IGI Global, Hershey, Pennsylvania, pp. 22–43.

Azmi, E., Che Rose, R.A., Awang, A. and Abas, A. (2023) Innovative and competitive: A systematic literature review on new tourism destinations and products for tourism supply. *Sustainability* 15(2), 1187.

Bunghez, C.L. (2021) The emerging trend of niche tourism: Impact analysis. *Journal of Marketing Research and Case Studies* 134710.

Burton, C. and Rogerson, J.M. (2023) Stakeholder perspectives concerning the progress of South Africa's urban ecotourism destination. *GeoSport for Society* 18(1), 9–29.

Câmara, E., Pocinho, M., Agapito, D. and Jesus, S.N.D. (2022) Positive psychology, well-being, and mindfulness: A successful partnership towards the development of meaningful tourist experiences. *Journal of Tourism, Sustainability and Well-Being* 10(1), 21–38.

Chen, L.I.-L., Scott, N. and Benckendorff, P. (2017) Mindful tourist experiences: A Buddhist perspective. *Annals of Tourism Research* 64, 1–12.

Chhabra, A. (2022) Social media transforming tourist behavior. In: Arora, S. and Sharma, A. (eds) *Event Tourism in Asian Countries: Challenges and Prospects*, 1st edn. Apple Academic Press, New York.

Chong, K.L. (2020) The side effects of mass tourism: The voices of Bali Islanders. *Asia Pacific Journal of Tourism Research* 25(2), 157–169.

Dalgıç, T. and Leeuw, M. (1994) Niche marketing revisited: Concept, applications and some European cases. *European Journal of Marketing* 28(4), 39–55.

Davis, F.D., Bagozzi, R.P. and Warshaw, P.R. (1992) Extrinsic and intrinsic motivation to use computers in the workplace. *Journal of Applied Social Psychology* 22(14), 1111–1132.

de Bruyn, C., Ben Said, F., Meyer, N. and Soliman, M. (2023) Research in tourism sustainability: A comprehensive bibliometric analysis from 1990 to 2022. *Heliyon* 9(8), e18874.

de Freitas Coelho, M., de Sevilha Gosling, M. and de Almeida, A.S.A. (2018) Tourism experiences: Core processes of memorable trips. *Journal of Hospitality and Tourism Management* 37, 11–22.

Di-Clemente, E., Hernández-Mogollón, J.M. and Campón-Cerro, A.M. (2020) Tourists' involvement and memorable food-based experiences as new determinants of behavioural intentions towards typical products. *Current Issues in Tourism* 23(18), 2319–2332.

Dutt, C. and Ninov, I. (2016) Tourists' experiences of mindfulness in Dubai, United Arab Emirates (UAE). *Journal of Travel & Tourism Marketing* 33(8), 1195–1212.

Eck, T., Zhang, Y. and An, S. (2023) A study on the effect of authenticity on heritage tourists' mindful tourism experience: The case of the Forbidden City. *Sustainability* 15(10), 7756.

Gadamer, H.G. (1979) The problem of historical consciousness. In: Rabinow, P. and Sullivan, W.M. (eds) *Interpretive Social Science: A Reader*. University of California Press, Oakland, California, pp. 103–160.

Gisore, R. and Ogutu, H. (2015) *Sustainable Tourism in Africa: Standards as Essential Catalysts*. Sector Standardisation Needs Review #9-3. ARSO Central Secretariat, Nairobi.

Jaakkola, E. (2020) Designing conceptual articles: Four approaches. *AMS Review* 10(1–2), 18–26.

Kachniewska, M. (2014) Aspects of tourist behavior. *Anatolia* 25(2), 318–320.

Kang, M. and Gretzel, U. (2012) Effects of podcast tours on tourist experiences in a National Park. *Tourism Management* 33(2), 440–455.

Kollmuss, A. and Agyeman, J. (2002) Mind the gap: Why do people act environmentally and what are the barriers to pro-environmental behavior? *Environmental Education Research* 8(3), 239–260.

Kumar, S., Atri, K. and Shekhar, S. (2018) Niche tourism: Sustainability and marketing strategies. *Tourism Innovations* 8(1), 1–9.

Kuntai, S.T. (2020) Awareness and implementation of the critical factors for sustainable tourism – a case of Maasai Mara National Reserve in Kenya. Doctoral dissertation, The Strathmore University, Nairobi.

Langer, E. (1989) *Mindfulness*. Addison-Wesley/Addison Wesley Longman, Boston, Massachusetts.

Langer, E.J. (1992) Matters of mind: Mindfulness/mindlessness in perspective. *Consciousness and Cognition* 1(3), 289–305.

Lee, Y. and Kim, I. (2018) Investigating key innovation capabilities fostering visitors' mindfulness and its consequences in the food exposition environment. *Journal of Travel & Tourism Marketing* 35(6), 803–818.

Lissner, I. and Mayer, M. (2020) Tourists' willingness to pay for Blue flag's new ecolabel for sustainable boating: The case of whale-watching in Iceland. *Scandinavian Journal of Hospitality and Tourism* 20(4), 352–375.

Lu, J. and Nepal, S.K. (2009) Sustainable tourism research: An analysis of papers published in the Journal of Sustainable Tourism. *Journal of Sustainable Tourism* 17(1), 5–16.

Luttrell, A., Brinol, P. and Petty, R.E. (2014) Mindful versus mindless thinking and persuasion. In: Ngnoumen, C.T. and Langer, E.J. (eds) *The Wiley Blackwell Handbook of Mindfulness*. Wiley, Chichester, UK, pp. 258–278.

Manwa, H., Boemah-Dlamini, D. and Coetzee, E. (2017) Mindfulness in cultural and wildlife tourism in Southern Africa. *International Journal of Environmental & Science Education* 12(8), 1913–1923.

Mayer, M.G., Gedi, M.V.K. and Kumar, M.G. (2023) The effect of digitalization on sustainable tourism and hospitality management. *Management and Finance Bulletin* 1(2), 1–14.

Mell, I. and Whitten, M. (2021) Access to nature in a post covid-19 world: Opportunities for green infrastructure financing, distribution and equitability in urban planning. *International Journal of Environmental Research and Public Health* 18(4), 1527.

Mgonja, J.T., Sirima, A. and Mkumbo, P.J. (2017) A review of ecotourism in Tanzania: Magnitude, challenges, and prospects for sustainability. In: Backman, K. and Munanura, I.E. (eds) *Ecotourism in Sub-Saharan Africa: Thirty Years of Practice*. Routledge, London, pp. 170–183.

Mihajlović, I. (2020) The impact of socio-economic changes in tourism on the business specialization of travel agencies. *WSEAS Transactions on Business and Economics* 17(35), 345–367.

Moreira, C.O., Ferreira, R. and Santos, T. (2021) Smart tourism and local heritage: Phygital experiences and the development of geotourism routes. In: Oliveira, L., Amaro, A.C. and Melro, A. (eds) *Handbook of Research on Cultural Heritage and Its Impact on Territory Innovation and Development*. IGI Global, Hershey, Pennsylvania, pp. 206–232.

Moscardo, G. (1996) Mindful visitors: Heritage and tourism. *Annals of Tourism Research* 23(2), 376–397.

Moscardo, G. (2017) Exploring mindfulness and stories in tourist experiences. *International Journal of Culture, Tourism and Hospitality Research* 11(2), 111–124.

Moscardo, G. and Pearce, P.L. (1986) Visitor centres and environmental interpretation: An exploration of the relationships among visitor enjoyment, understanding and mindfulness. *Journal of Environmental Psychology* 6(2), 89–108.

Mutum, D.S. and Ghazali, E.M. (2023) *Consumers, Society and Marketing: A Sustainability Perspective*. Springer Nature, Cham, Switzerland.

Nel, E. and Rossouw, R. (2020) The role of niche tourism in achieving sustainable development goals: A South African perspective. *Journal of Sustainable Tourism* 28(9), 1295–1314.

Novelli, M. (2018) Niche tourism: Past, present and future. In: Cooper, C., Volo, S., Gartner, W.C. and Scott, N. (eds) *The SAGE Handbook of Tourism Management: Applications of Theories and Concepts to Tourism*. SAGE Publications Ltd, Melbourne, Australia, pp. 344–359.

Ntshingila, L. and Musakwa, W. (2020) Community-based tourism enterprises in South Africa: Promoting sustainable niche tourism. *Journal of Ecotourism* 19(2), 156–173.

O'Dell, T. and Billing, P. (2005) Experiencescapes: Tourism, culture and economy. *Tourist Studies* 5(1), 107–109.

Pearce, P.L. (2005) *Tourist Behaviour: Themes and Conceptual Issues*. Channel View Publications, Clevedon, UK.

Poon, A. (1989) Consumer strategies for a new tourism. In: Cooper, C. (ed.) *Progress in Tourism, Recreation and Hospitality Management*. Bellhaven, London, pp. 91–102.

Poria, Y., Butler, R. and Airey, D. (2004) Links between tourists, heritage, and reasons for visiting heritage sites. *Journal of Travel Research* 43(1), 19–28.

Proos, E. and Hattingh, J. (2022) Dark tourism: Growth potential of niche tourism in the Free State Province, South Africa. *Development Southern Africa* 39(3), 303–320. DOI: 10.1080/0376835X.2020.1847636.

Ramchander, P. (2007) Township tourism – blessing or blight? The case of Soweto in South Africa. In: Richards, G. (ed.) *Cultural Tourism: Global and Local Perspectives*. Routledge, Binghamton, New York, pp. 39–68.

Richards, G. (2021) Rethinking niche tourism: The example of backpacking. *Croatian Regional Development Journal* 2(1), 1–10.

Rogerson, C.M. (2020) Niche tourism and regional development in South Africa. *Africa Insight* 50(2), 90–105.

Rogerson, C.M. and Rogerson, J.M. (2021a) Niche tourism research and policy: International and South African debates. *African Journal of Hospitality, Tourism and Leisure* 10(4), 1131–1151.

Rogerson, C.M. and Rogerson, J.M. (2021b) Niche tourism in South Africa: A renewed policy focus in the covid-19 environment. *GeoJournal of Tourism and Geosites* 39(4), 1379–1387.

Rubin, S.D., Lee, W., Paris, C.M. and Teye, V. (2016) The influence of mindfulness on tourists' emotions, satisfaction and destination loyalty in Fiji. *Travel and Tourism Research Association: Advancing Tourism Research Globally* 54.

Sarantou, M., Kugapi, O. and Huhmarniemi, M. (2021) Context mapping for creative tourism. *Annals of Tourism Research* 86, 103064.

Sekole, M.Q., Lethoko, M., Mautjana, M.H. and Makombe, G. (2022) Community perceptions on tourism development in South Africa: The case of developing a tourism route in the City of Mbombela Local Municipality, Mpumalanga Province, South Africa. *Local Economy* 37(6), 462–480.

Sert, A.N. (2017) Niche marketing and tourism. *Journal of Business Management and Economic Research* 1(1), 14–25.

Shapiro, S. (2020) *Rewire Your Mind: Discover the Science and Practice of Mindfulness*. Hachette, Edinburgh, UK.

Sheldon, P.J. (2020) Designing tourism experiences for inner transformation. *Annals of Tourism Research* 83, 102935.

Singh, T. (2021) Cultural heritage and sustainable niche tourism in South Africa. *Tourism Review International* 25(1), 47–63.

Somrak, W. and Khaewmesuan, R. (2022) Tourism consumption trends changing and tourism promotion after the outbreak of Covid-19. *ASEAN Journal on Hospitality and Tourism* 20(3), 1–13.

Stankov, U. and Filimonau, V. (2023) Here and now – the role of mindfulness in post-pandemic tourism. *Tourism Geographies* 25(1), 374–389.

Stankov, U., Gretzel, U. and Filimonau, V. (2022) *The Mindful Tourist: The Power of Presence in Tourism*. Emerald Publishing Limited, Bingley, UK.

Steg, L. and Vlek, C. (2009) Encouraging pro-environmental behaviour: An integrative review and research agenda. *Journal of Environmental Psychology* 29(3), 309–317.

Suryandari, N. and Rahmawati, F.N. (2022) Mindful tourism: Management of tourist destinations as an effort to counter stereotypes of the Madurese people. In: Pambuko, Z.B. (ed.) *4th Borobudur International Symposium on Humanities and Social Science*, Atlantis Press, pp. 186–195.

Taylor, L.L. and Norman, W.C. (2019) The influence of mindfulness during the travel anticipation phase. *Tourism Recreation Research* 44(1), 76–90.

United Nations (2015) Global Sustainable Development Report: 2015 Edition, Advance Unedited Version. Available at: https://sustainabledevelopment.un.org/content/documents/1758GSDR%202015%20Advance%20Unedited%20Version.pdf (accessed 22 October 2024).

United Nations (2018) Transforming Our World: The 2020 Agenda for Sustainable Development. Available at: https://sustainabledevelopment.un.org/content/documents/21252030%20Agenda%20for%20Sustainable%20Development%20web.pdf (accessed 31 March 2024).

Van Niekerk, D. (2022) Exploring niche tourism opportunities in South Africa: A case study of heritage and adventure tourism. *Journal of Heritage Tourism* 17(1), 25–42.

Vos, J. (2015) Township tourism plays a vital role. Available at: http://www.tourismupdate.co.za/article/124789/Township-tourism-plays-a-vital-role (accessed 25 March 2024).

Walls, A.R., Okumus, F., Wang, Y.R. and Kwun, D.J.W. (2011) An epistemological view of consumer experiences. *International Journal of Hospitality Management* 30(1), 10–21.

Wang, Y.C., Chen, P.J., Shi, H. and Shi, W. (2021) Travel for mindfulness through Zen retreat experience: A case study at Donghua Zen Temple. *Tourism Management* 83, 104211.

Weber, F. (2019) Demand for sustainable tourism. In: Lund-Durlacher, D., Dinica, V., Reiser, D. and Fifk, M.S. (eds) *Corporate Sustainability and Responsibility in Tourism: A Transformative Concept*. Springer Nature, Cham, Switzerland, pp. 265–281.

Zhang, Y., Woo Park, Y. and Cole, S.T. (2016) Factors influential to tourism experiences of people with mobility impairments. *Travel and Tourism Research Association: Advancing Tourism Research Globally* 25.

2 Exploring Visitors' Experiences at the Heritage Site of Sydney's Quarantine Station

Feras Orekat*, Glenn Murray and Andrew Le
International College of Management, Sydney, Australia

Abstract

Comprehending visitors' perspectives is essential for heritage sites striving to deliver meaningful experiences. This chapter describes a qualitative study that explores visitors' encounters at the heritage site of Sydney's Quarantine Station (Q Station). Through an analysis of 581 visitor book comments spanning from 2017 to 2023, the preliminary findings reveal three primary factors influencing on-site visitors: cultural aspects (site authenticity, historical significance and visitors' cultural connections), environmental elements (wildlife and scenic vistas) and secular amenities (service quality, food and accommodation). These factors align with Falassi's (1987) model of 'space as time out of time', which highlights the interplay between cultural, natural and secular dimensions in shaping visitor experiences. Among these factors, the site's cultural characteristics emerge as the most influential in shaping visitors' overall experiences. Notably, visitors with prior cultural, ancestral or personal connections to the site report more memorable experiences than those without such ties. However, the study also unveiled dissatisfaction among Aboriginal Australian visitors, who felt that the site's management inadequately represented their indigenous culture. Based on these findings, the research recommends that site managers prioritize the conservation of the Q Station's cultural characteristics while incorporating a more comprehensive representation of Australian Aboriginal culture. Simultaneously, maintaining and enhancing the environmental and secular amenities should not be neglected, as these elements contribute to the holistic visitor experience.

Keywords: Visitor experience, Q Station, heritage site, ancestral connection

Introduction

Conceptualizing visitors' experiences at heritage sites is a multifaceted endeavour involving diverse perspectives and dimensions. While some scholars emphasize the emotional and mental aspects of the visitor experience (Volo, 2009; Walls *et al.*, 2011; Leri and Theodoridis, 2019; Hosany *et al.*, 2020), others underscore the role of site offerings, activities and interpretation (Csikszentmihalyi and Robinson, 1990; Schmitt, 1999; Selstad, 2007; Kuo *et al.*, 2018). In this regard, Packer and Ballantyne (2016, p. 128) argued that 'although commonly considered the core product offered by visitor attractions such as museums, zoos, and heritage sites, the visitor experience has proved a difficult construct to both define and measure'. They presented multiple facets of the visitor experience and classified them into four main groups: (i) experience as a flow of consciousness, with this perspective being rooted in anthropology,

*Corresponding author: forekat@icms.edu.au

 Niche Tourism and Sustainability: Perspectives, Practices and Prospects (eds A. Farmaki *et al.*)
DOI: 10.1079/9781800626669.0002

perceiving the visitor experience as a flow of consciousness (Bruner, 1986); (ii) experience as a subjective response to an event or stimulus (Schmitt, 1999); (iii) experience as a memorable impression (Carbone and Haeckel, 1994); and (iv) experience as a designed or staged offering (Carù and Cova, 2003; Volo, 2009).

Additionally, cultural heritage sites pose a unique challenge, as they represent multiple perspectives, stories and memories of specific groups (Lehto *et al.*, 2004; Chen and Rahman, 2018), holding immense appeal for visitors seeking cultural connections. Heritage tourism offers a unique opportunity for visitors to connect with the past, value diverse cultural stories and deepen their understanding of the world's heritage (Lee and Chhabra, 2015). It serves as a bridge between the past and present, fostering an appreciation for preserving historical and cultural legacies. Additionally, sustainability is essential for heritage sites to mitigate their negative impacts on the environment and society while encouraging practices that support responsible tourism (Fyall and Garrod, 1998; Kuo, 2002; Chang and Horng, 2010; Ramzy and Morsi, 2020).

Sustainability in niche heritage sites like the Quarantine Station involves balancing the three Ps (people, planet, and profit) (Molina and Rajagopal, 2023) and four pillars of sustainability (social, economic, cultural, and environmental) (Nurse, 2006) to preserve sites for future generations while meeting current needs. The site demonstrates social sustainability through community engagement, involving local residents, national parks and government in preservation and management. Educational programmes and accessibility initiatives promote inclusivity and raise awareness about the site's history and the importance of sustainability. Culturally, the site maintains authenticity by preserving original buildings, materials and construction techniques. Documentation of the site's history and cultural significance aids preservation efforts. Environmental sustainability is achieved through protecting local flora and fauna, energy-efficient systems and a waste management programme, including composting and plastic reduction. These efforts enhance the visitor experience while reducing the site's carbon footprint.

Sydney's Quarantine Station (Q Station) is a unique heritage site embodying the intertwined narratives of indigenous and colonial Australia (Longhurst, 2016). As the first site where early European settlers isolated sick individuals, the Q Station holds immense historical and cultural significance (Hinkson, 2003). However, despite its importance, there is a dearth of research exploring visitors' experiences at this remarkable site. The aim of this chapter is to explore visitors' experiences at the niche heritage site of Sydney's Quarantine Station to deliver meaningful experiences.

Literature Review

Tourism venues have long been concerned with visitor experience and satisfaction; a task that becomes more complex when visitors' expectations vary according to their reasons for visiting (Damanik and Yusuf, 2022). At heritage sites, the changing generational demands for interactive experiences pose a challenge for site managers to adapt. According to Sharkawi and El-Rahman (2022), the relationship between visitors and the intangible cultural heritage elements at these sites should be interactive. They suggest that site managers should move away from traditional interpretation methods that rely on 'boring tour guides' to explain the site's story. Instead, an interactive interpretation approach can enhance the visitor experience and motivate repeat visits.

In their attempt to understand the importance of interactive interpretation, Viljoen and Kruger (2020) argued that it has not yet been widely recognized in the literature as a distinct form of interpretation. They posited that the application of technology, such as touchscreens and interactive puzzles, to educate visitors can help raise awareness about conservation issues at destinations. Through these interactive elements, visitors are encouraged to actively engage with the content, potentially leading to greater retention and connection to the conservation messaging.

Conceptualizing the visitor experience

Dewey, (1938, p. 43) defines a visitor's experience as a 'transaction between an individual and what, at the time, constitutes his environment'. Similarly, Packer and Ballantyne (2016) emphasize the experience as a specific response to multifaceted external and staged activities, settings or events. These staged activities should provide opportunities for engagement, cross-domain interactions, and exposure to the richness of objects and materials (Holbrook and Hirschman, 1982; Cohen, 1996; Sharkawi and El-Rahman, 2022). Holistically, Volo (2009) focuses on the 'experience essence' in the visitor's mind and the experience offering created by the destination's management team.

Walls *et al.* (2011, p. 11) suggested that tourism providers need to 'orchestrate only the opportunity for an experience'. In this sense, providers should create an 'atmosphere' where visitors can receive a transformative experience, creating value and encouraging repeat visits. Schmitt (1999) argued that although receiving a transformative experience is subjective, site offerings enhance the visitor experience by providing the 'right environment and setting for the desired customer to emerge' (p. 60). Swarbrook and Horner (2011) also highlighted the importance of tapping into visitors' emotions by creating activities that suit their lifestyles and needs.

In his attempt to understand how visitors perceive time in festivals and events, Falassi (1987) developed the theoretical model of 'space as time out of time'. He argued that visitors perceive time differently when visiting a tourist destination due to internal and external factors such as socialization with locals, culture and previous knowledge about the site. Sites of heritage importance, particularly those with religious or sacred connotations, offer visitors an opportunity to mentally recharge, break away from their normal daily patterns and engage with spiritual dimensions that may be lacking in their typical surroundings. Falassi's model suggests that the experiences visitors have at these heritage locations are shaped by their socialization, underlying motivations and prior knowledge or understandings they bring with them. This model was validated by Bond (2013), focusing on visitors' experiences at English cathedrals, and Orekat (2015), on visitors' experiences at sites related to monotheistic religions. Their findings (Bond, 2013; Orekat, 2015) corroborate Falassi's (1987) argument that heritage sites with rich history significantly impact the visitor experience by disconnecting them from the mundane world and encouraging spiritual connection to the site.

Visitor experience from a managerial perspective

Adopting a managerial lens, Pine II and Gilmore (1999) developed the 'Experience Economy' concept, stating that valuable experiences should encompass one or more of the four Es: entertaining, educational, escapist and aesthetic. They argued that experiences should be transformative and memorable and help visitors become better versions of themselves. Similarly, Falk and Dierking (2013) referred to the totality of the experience, including anticipated and remembered aspects, as part of the visitor experience. Hence, visitors are not passive participants but active learners eager to immerse themselves in a transcendent experience with psychological value (Selstad, 2007).

An integrated approach combining robust operational systems, personnel management and comprehending visitor needs is crucial for managing the visitor experience at a heritage site (Cohen, 1996; Orekat, 2015). Clark (1989, p. 92) contended that site managers must 'understand how to oversee all aspects of their offerings, not just the quality of physical products, but also supporting materials, delivery methods, as well as all interactions with customers during daily operations'. This implies that services involving human elements, such as heritage tourism, necessitate an integrated management strategy encompassing the offering, the customer and the service team responsible for delivery (Packer and Ballantyne, 2016).

Carù and Cova (2003, p. 270) stated that an experience comprises visitors' emotions at the site and ongoing activities that can be translated into a 'reportable story'. Storytelling shapes the destination's brand and creates an emotional connection between visitors and the destination (Singer *et al.*, 2011; Akgün *et al.*, 2015). Through storytelling, visitors can have a nostalgic experience by reminiscing about previous encounters. Hence, 'when visitors record or communicate responses to their visit through narration, photography, or social media, it becomes, for them, a memorable experience' (Packer and Ballantyne, 2016, p. 133). Perhaps one aim of interpretation should be to furnish visitors with tools and resources that enable them to construct their meaning and create their captivating narratives, thereby making the experience 'extraordinary' or 'remarkable'.

Whether adopting an internal approach, where visitor experience is defined by subjective and personal factors, or an external approach, where experiences span an extended time (Dewey, 1938; Cutler and Carmichael, 2010; Hennes, 2010; Walls *et al.*, 2011), it is apparent that a visitor experience is subject to diverse internal and external factors that need to be studied carefully and in the specific context. This is because heritage sites have unique stories that can affect each visitor differently. Storytelling, site activities, settings and subjective factors interplay while a visitor is at a heritage site, and there is a need to investigate these components in more depth to contribute to the conceptualization of the visitor experience (Singer *et al.*, 2011; Akgün *et al.*, 2015; Packer and Ballantyne, 2016).

Case Study: From Quarantine to Q Station

The Q Station on Sydney's North Head is listed on the National Heritage Register as a site of outstanding heritage and cultural value to the nation. The land of North Head was used by indigenous peoples long before the arrival of colonial settlers in Sydney Harbour in 1788. Jean Foley's research (*Healings and Burials at Sydney's North Head*) provides a narrative of the site's significance and use by Aboriginal communities (Foley, 2019). In *Repossession of Our Spirit*, Dennis Foley (2001) also highlights North Head as an important country for the Guringai people, identifying it as a sacred place where the Koradgee, or wise ones, taught and healed.

From 1828 to 1984, an operational maritime quarantine station occupied the land of North Head adjacent to Spring Cove. The historical evidence on the site today includes 67 buildings and a permanent moveable heritage collection, including both primary and secondary historical sources. Together, these tell varied stories – the quarantine process over time, personal accounts of migration from different periods and cultures, the Australian federation, World War I, Australian immigration policy including the Immigration Restriction Act 1901, migrant detention, Operation Babylift and changing knowledge of health, vaccines and the treatment of infectious diseases (Hobbins *et al.*, 2016).

The Q Station's history is deeply intertwined with the narratives of both indigenous Australians and colonial settlers. While serving as a quarantine station for incoming migrants and travellers, it also held profound spiritual significance for the Guringai people as a sacred site for teaching and healing. This duality of purpose and meaning makes the Q Station a unique heritage site, embodying the complex histories and cultural intersections that have shaped modern Australia.

Q station's contemporary management and operations

In 2006, 22 years after the operational quarantine station ceased operations, Mawland secured a leased agreement with conditions to establish a sustainable tourism venture within the site premises. The land remains under the ownership of the New South Wales state government as part of the Sydney Harbour National Park. Mawland's plan involved the adaptive reuse of the site and its existing structures, resulting in a business comprising 106 accommodation rooms, a restaurant, conference facilities, event spaces, educational programmes and guided tours.

In 2022, a business transaction transferred the leasehold and operations to North Head Operations. As a medium-sized enterprise employing around 100 personnel, North Head Operations continues to operate the Q Station as a tourist destination. Table 2.1 indicates that between 1 October 2021 and 31 December 2023, the Q Station hosted 53,138 visitor arrivals (M. Pedro, manager of front office at Q Station, Sydney, 6 August 2023, personal communication), given that the number of tourist arrivals was significantly impacted by the COVID-19 pandemic.

Table 2.1. Number of arrivals to Q Station between the years 2021 and 2023. (M. Pedro, manager of front office at Q Station, Sydney, 6 August 2023, personal communication.)

Year	Tourist arrivals number
2021 (1 October–31 December)	6776
2022	15,834
2023	30,528

The contemporary management of the Q Station concentrates on striking a balance between its historical importance and sustainable tourism practices. The adaptive reuse of existing structures preserves the site's architectural heritage while repurposing them for modern visitor accommodation, dining and event facilities. Educational programmes and tours aim to impart the rich history of the Q Station to visitors, fostering an appreciation for its cultural and national significance.

Maintaining this delicate equilibrium between preservation and tourism presents ongoing challenges for site managers. Ensuring the site's authenticity while meeting modern visitors' evolving needs and expectations requires meticulous planning and execution. Collaboration with stakeholders, including indigenous communities and historical societies, is crucial to upholding the Q Station's multifaceted narratives and cultural significance.

Methods

This research utilized a qualitative methodology by examining 581 comments from visitor comment books spanning 2017 to 2022 (Isaac and Budryte-Ausiejiene, 2015). A triple-coding technique was employed to scrutinize comments written in English, following the grounded theory approach (Charmaz, 2006; Glaser and Strauss, 2017). Initially, themes were openly coded to pinpoint substantive remarks; then axial coding was used to classify them into three primary categories. Finally, selective coding uncovered the interconnectedness among the identified themes and categories (Creswell, 2013; Sotiriadou *et al.*, 2014).

The open coding phase involved a line-by-line analysis of the comments, identifying and labelling concepts and phenomena present in the data (Strauss and Corbin, 1998). This process allowed for the emergence of initial codes that captured the essence of visitors' experiences and perceptions. During the axial coding, relationships between these initial codes were explored, developing broader categories and subcategories (Corbin and Strauss, 2014).

Through the rigorous application of this triple-coding process, the study ensured a comprehensive and systematic analysis of visitors' feedback, offering deep insight into their experiences at the Q Station. This methodological approach not only bolstered the credibility and reliability of the results but also facilitated the detection of nuanced themes and patterns that might have been overlooked in a less meticulous analysis (Ayton *et al.*, 2023).

The analysis revealed three key themes emerging from the visitor comments: (i) the *cultural* theme encompassed aspects related to the site's historical significance, heritage value and cultural interpretation. Visitors' reflections on the quarantine station's role in Australia's immigration history and connection to diverse cultural narratives were prominent; (ii) the *secular* theme focused on the non-religious aspects of visitors' experiences, including educational value, leisure activities and general site amenities. Comments related to guided tours, interactive displays and the overall visitor experience were categorized under this theme; and (iii) the *environmental* theme captured visitors' perceptions and experiences related to the natural setting of the Q Station. Comments on the site's landscape, flora and fauna, and conservation efforts were analysed under this category.

The selective coding phase allowed for exploring relationships between these themes, revealing how cultural, secular and environmental aspects often intertwine in visitors' experiences. This integrative approach provided a holistic understanding of the multifaceted nature of heritage tourism at the Q Station. By employing this comprehensive analytical framework, the study captured the complexity of visitors' experiences and perceptions, offering valuable insights for heritage management and tourism development at the Q Station and similar sites.

Findings and Discussion

The findings suggest that three primary elements play a significant role in shaping the experiences of onsite visitors: cultural factors (including the authenticity and historical importance of the site as well as visitors' cultural ties), environmental features (such as wildlife and picturesque views) and secular factors (like service quality, dining options and accommodation).

Cultural characteristics

The influence of visitors' cultural connections on shaping their experiences is evident from 93 statements. Sam described the site as 'a fascinating and beautiful place to visit. The tours were very informative, and I would love to visit again'. Ambrose stated it is 'an exciting place and would love to stay here 1 day to learn the history'.

However, the site's story appeals more to those with ancestral connections regarding repeat visitation and engagement. Jack, with an ancestral connection, repeated visits several times, saying, 'Third visit here. Family connection. Father R.L. Challinor was a ship engineer on the 'Aorangi'. Great memories returning of a wonderful man'. Ian also said, 'I have visited Q Station many times; my grandfather came and started a new life here. The Q Station is an exceptional site for me because it connects me with my ancestors'. Rose described her strong emotion: 'I cannot explain how I feel today! My grandfather told me his father came through Q Station, and today I saw my great-grandfather's name in the records at the tour desk, to see my great-grandfather's name put tears in my eyes, can't imagine what he experienced while staying here'.

Some stressed preserving the site's history and artefacts. Sarah stated, 'This is such an amazing piece of history. I am so thrilled it has been preserved for future generations'. Bao said, 'As an Australian man with Chinese heritage, although I live 5 km away, I did not know this site has a rich Chinese history. I learned that Chinese immigrants came here during the gold rush and sadly many suffered because of their ethnicity. Definitely will bring my family and come back, and hope the site remains without adding modern elements'.

However, some felt the story focuses on 'white European heritage' without showcasing indigenous heritage enough. Terry said disappointedly, 'Shame our indigenous heritage is not acknowledged in any way here! (our country's real past)'. Henley also commented, 'the heritage is vital for preservation. Congratulations! Some indigenous tours would be informative and respectful'.

The intangible elements, such as ancestral connections and historical significance represented in the site's story, appear to be the most influential factors affecting visitors' quarantine station experiences. This supports arguments by Carù and Cova (2003), Singer *et al.* (2011), Akgün *et al.* (2015) and Packer and Ballantyne (2016) that site storytelling significantly impacts visitors' heritage site experiences.

Secular characteristics

The impact of the site's services (food, accommodation) is shown in 23 quotes. Jenny stated, 'So cool! Love the café and the accommodation'. Sam said, 'What a place to visit, from accommodation to food, ghost tour, but best of all, the beautiful staff, awesome'. Brad suggested, 'I suggest advertising more about the site's restaurant and cafe, such a beautiful experience to have dinner with family in an old building with a gorgeous view'.

However, 18 of the 23 quotes mentioned the site's authenticity and museum. Edward said, 'What an authentic place! Great food, accommodation, and museum. Thanks for keeping it for our future generations'. Daniel commented, 'I found the museum remarkably interesting to walk around. I would like to return in the future to this great spot'. Rose said, 'Stunning views, well-kept with beautiful staff, all willing to help. History is amazing; I feel so lucky for where I live and have been'. Mark highlighted the unique dining experience: 'The fact that we are dining in a 100 + year old building made our experience very unique. I prefer to eat here more than going to a 5-star restaurant, at least you don't get this authentic experience over there'.

These responses support studies highlighting secular characteristics' importance, like good food, accommodation, service and so on, in maximizing the visitor experience at heritage sites (Agyeiwaah *et al.*, 2016; Liping and Alaedini, 2018). However, the site's history made the dining experience more joyful. Cohen (2006) recognizes sustaining authenticity's importance in visitors' experiences. This aligns with Cohen's (2006) contention that authentic heritage sites exert a cognitive influence on visitors, potentially manifesting as increased knowledge, affective impacts (e.g. heightened feelings of attachment or commitment), behavioural effects (e.g. engaging in religious rituals, involvement with the local community) or a desire to revisit the site.

Environmental characteristics

Out of the 581 comments, 41 were related to Q Station's wildlife and views. As a national park with rare animals like bandicoots and echidnas, the wildlife enhanced visitors' experience. Sarah said, 'Had a relaxing stay and loved the Q Station and ghost tour and enjoyed the proximity to wildlife. The food in the restaurant was excellent, and all staff were helpful'. Zack described encountering an echidna: 'I met an echidna today! Such a lovely family day trip'. Rose commented, 'Great croissant and coffee'. William said, 'Had a very relaxing stay. We loved the Q Station and ghost tour and enjoyed the proximity to the wildlife. The food in the restaurant is excellent, and all staff here are friendly and helpful. Views spectacular'. Highlighting preserving the wildlife's importance, Pedro said, 'This site is full of wild animals, in the morning I saw a seal near the beach, and later an echidna, and at night a bandicoot. I'm a very lucky man and what an amazing site, wish the people in charge look after the animals because they are awesome'.

These responses support environmental conservation's importance for visitor satisfaction at heritage sites (MacCannell, 2011), considering sightseeing connects visitors physically and emotionally with nature, heritage and others. This confirms conserving environmental aspects' necessity for visitor satisfaction.

According to Relph (1976, p. 142), a place combines 'setting, landscape, ritual, people, personal experiences, care and concern for home, and context of other places'. Fundamentally, visitors are the customers/users of the tourism venue and services. The provider is responsible for meeting satisfaction and reflecting expectations. The findings echo satisfaction defined by Della Corte *et al.* (2015, p. 38) as 'the overall feeling/attitude accompanying a product's purchase'. Crucially, visitor satisfaction with the experience and services offered builds upon accumulated feelings, emotions and fulfilment. Heritage sites incorporate multidimensional cultural, environmental and secular aspects.

The findings support Falassi's (1987) model of 'space as time out of time', contending that sites with historical and spiritual values give visitors time for psychological refreshment. Fig. 2.1 conceptualizes the visitor experience at Q Station, with three interconnected characteristics influencing it: cultural, environmental and secular. Visitors interested in secular aspects (service, food, accommodation) and environmental attributes (wildlife and views) were also interested in the cultural elements of the site's story and authenticity. However, visitors with ancestral connections appeared more attached, repeating visits more than other visitor types. Four comments mentioned indigenous Australian culture was not adequately represented.

Conclusion and Implications

This chapter described research investigating the factors impacting visitors' experiences at the Q Station heritage site. Data were extracted from visitor book comments at the Q Station between 2017 and 2023. Three main characteristics emerged as highly important: cultural (site's authenticity, historical significance, and visitors' cultural connections), environmental (wildlife and views) and secular (service quality, food and accommodation). However, the cultural values provoked visitors' emotions more profoundly than the other characteristics.

The research corroborates Falassi's (1987) theoretical model for understanding the visitor experience at the Q Station. The findings indicate

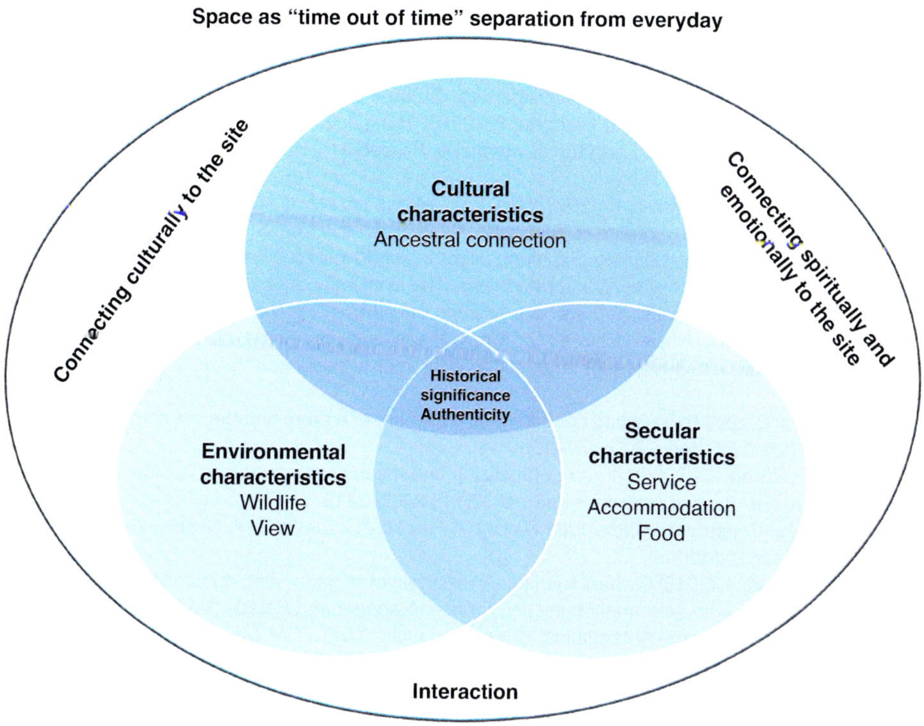

Fig. 2.1. A model conceptualizes the visitor experience at the Q Station.

that visitors with ancestral or cultural ties to the site had a more memorable experience than those without such connections. Additionally, most visitors underwent a transformative experience at the site, attributed to its unique narrative and rich historical background. Based on these findings, the research recommends that site managers prioritize preserving the Q Station's cultural characteristics, as they are the main factors shaping the visitor experience. Out of the 581 comments analysed, only four were related to Aboriginal Australians, and all these comments suggested more representation of indigenous culture. Therefore, the study recommends designing a tour focusing on Aboriginal Australian culture and further research investigating their experiences at the Q Station. Incorporating a more comprehensive representation of narratives and perspectives would enrich the visitor experience and contribute to preserving and celebrating the site's multifaceted cultural heritage.

Furthermore, Q Station staff demonstrate commitment to site preservation, fostering local engagement and stewardship. Educational programmes inform visitors about the site's history and sustainability importance, supporting responsible tourism. These initiatives benefit the local community and businesses economically through tourism and related activities, contributing to societal well-being while improving the sustainability of destinations. Q Station's management practices serve as a model for other heritage sites in terms of community engagement and sustainability, further benefiting the local economy through increased tourism.

References

Agyeiwaah, E., Adongo, R., Dimache, A. and Wondirad, A. (2016) Make a customer, not a sale: Tourist satisfaction in Hong Kong. *Tourism Management* 57, 68–79.

Akgün, A.E., Keskin, H., Ayar, H. and Erdoğan, E. (2015) The influence of storytelling approach in travel writings on readers' empathy and travel intentions. *Procedia – Social and Behavioral Sciences* 207, 577–586.

Ayton, D., Tsindos, T. and Berkovic, D. (2023) *Qualitative Research: A Practical Guide for Health and Social Care Researchers and Practitioners*. Monash University, Melbourne, Australia.

Bond, N. (2013) Searching for meaning: Exploring pilgrimage and religious heritage tourism experiences in English religious heritage sites. Doctorate thesis, The University of Queensland, Brisbane, Australia.

Bruner, E.M. (1986) Experience and its expressions. In: Turner, V.W. and Bruner, E.M. (eds) *The Anthropology of Experience*. University of Illinois Press, Urbana, Illinois, pp. 3–30.

Carbone, L.P. and Haeckel, S.H. (1994) Engineering customer experiences. *Marketing Management* 3(3), 8–19.

Carù, A. and Cova, B. (2003) Revisiting consumption experience: A more humble but complete view of the concept. *Marketing Theory* 3(2), 267–286.

Chang, T.-Y. and Horng, S.-C. (2010) Conceptualizing and measuring experience quality: The customer's perspective. *The Service Industries Journal* 30(14), 2401–2419.

Charmaz, K. (2006) *Constructing Grounded Theory: A Practical Guide through Qualitative Analysis*. SAGE Publications Ltd, London.

Chen, H. and Rahman, I. (2018) Cultural tourism: An analysis of engagement, cultural contact, memorable tourism experience and destination loyalty. *Tourism Management Perspectives* 26, 153–163.

Clark, G. (1989) Managing the 'intangibles' of service quality. *The TQM Magazine* 1(2). DOI: 10.1108/EUM0000000002994.

Cohen, E.H. (2006) Research in religious education: Content and methods for the postmodern and global era. *Religious Education* 101(2), 147–152.

Cohen, E.H. (1996) A phenomenology of tourist experiences. In: Apostolopoulos, Y., Leivadi, S. and Yiannakis, A. (eds) *The Sociology of Tourism: Theoretical and Empirical Investigations*. Routledge, London, pp. 56–71.

Corbin, J. and Strauss, A. (2014) *Basics of Qualitative Research: Techniques and Procedures for Developing Grounded Theory*. SAGE Publications, Inc, Thousand Oaks, California.

Creswell, J. (2013) *Research Design: Qualitative, Quantitative, and Mixed Methods Approaches*. SAGE Publications, Inc, Thousand Oaks, California.

Csikszentmihalyi, M. and Robinson, R.E. (1990) *The Art of Seeing: An Interpretation of the Aesthetic Encounter*. J.P. Getty Museum, Los Angeles, California.

Cutler, S.Q. and Carmichael, B.A. (2010) The dimensions of the tourist experience. In: Morgan, M., Lugosi, P. and Ritchie, J.R.B. (eds) *The Tourism and Leisure Experience: Consumer and Managerial Perspectives*. Channel View Publications, Bristol, UK, pp. 3–26.

Damanik, J. and Yusuf, M. (2022) Effects of perceived value, expectation, visitor management, and visitor satisfaction on revisit intention to Borobudur temple, Indonesia. *Journal of Heritage Tourism* 17(2), 174–189.

Della Corte, V., Sciarelli, M., Cascella, C. and Del Gaudio, G. (2015) Customer satisfaction in tourist destination: The case of tourism offer in the City of Naples. *Journal of Investment and Management* 4(1–1), 39–50.

Dewey, J. (1938) *Experience and Education*. Collier-Macmillan, London.

Falassi, A. (1987) *Time Out of Time: Essays on the Festival*. University of New Mexico Press, Albuquerque, New Mexico.

Falk, J.H. and Dierking, L.D. (2013) *The Museum Experience Revisited*. Left Coast Press, Inc, Walnut Creek, California.

Foley, D. (2001) *Repossession of Our Spirit: Traditional Owners of Northern Sydney*. ANU Press, Canberra.

Foley, J.D. (2019) *Healings and Burials at Sydney's North Head: Aboriginal People c.1788, Quarantined People 1835–1924*. Literary Dramatic Musical, Canberra.

Fyall, A. and Garrod, B. (1998) Heritage tourism: At what price? *Managing Leisure* 3(4), 213–228.

Glaser, B. and Strauss, A. (2017) *Discovery of Grounded Theory: Strategies for Qualitative Research*. Routledge, New York.

Hennes, T. (2010) Exhibitions: From a perspective of encounter. *Curator* 53(1), 21–33.
Hinkson, M. (2003) Encounters with aboriginal sites in metropolitan Sydney: A broadening horizon for cultural tourism? *Journal of Sustainable Tourism* 11(4), 295–306.
Hobbins, P., Frederick, U.K. and Clarke, A. (2016) *Stories from the Sandstone: Quarantine Inscriptions from Australia's Migrant Past*. Arbon Publishing, Sydney, Australia.
Holbrook, M.B. and Hirschman, E.C. (1982) The experiential aspects of consumption: Consumer fantasies, feelings, and fun. *Journal of Consumer Research* 9(2), 132–140.
Hosany, S., Hunter-Jones, P. and McCabe, S. (2020) Emotions in tourist experiences: Advancing our conceptual, methodological and empirical understanding. *Journal of Destination Marketing & Management* 16, 100444.
Isaac, R.K. and Budryte-Ausiejiene, L. (2015) Interpreting the emotions of visitors: A study of visitor comment books at the Grūtas Park Museum, Lithuania. *Scandinavian Journal of Hospitality and Tourism* 15(4), 400–424.
Kuo, I.L. (2002) The effectiveness of environmental interpretation at resource-sensitive tourism destinations. *International Journal of Tourism Research* 4(2), 87–101.
Kuo, N.-T., Cheng, Y.-S., Chang, K.-C. and Hu, S.-M. (2018) Assessing the asymmetric impact of interpretation environment service quality on museum visitor experience and post-visit behavioral intentions: A case study of the National Palace Museum. *Asia Pacific Journal of Tourism Research* 23(7), 714–733.
Lee, W. and Chhabra, D. (2015) Heritage hotels and historic lodging: Perspectives on experiential marketing and sustainable culture. *Journal of Heritage Tourism* 10(2), 103–110.
Lehto, X.Y., Cai, L.A., O'Leary, J.T. and Huan, T.-C. (2004) Tourist shopping preferences and expenditure behaviours: The case of the Taiwanese outbound market. *Journal of Vacation Marketing* 10(4), 320–332.
Leri, I. and Theodoridis, P. (2019) The effects of the winery visitor experience on emotions, satisfaction and on post-visit behaviour intentions. *Tourism Review* 74(3), 480–502.
Liping, C.A. and Alaedini, P. (eds) (2018) *Quality services and experiences in hospitality and tourism. Bridging Tourism Theory and Practice*, Vol. 9. Emerald Publishing Limited, Bingley, UK.
Longhurst, P. (2016) Quarantine matters: Colonial quarantine at North Head, Sydney and its material and ideological ruins. *International Journal of Historical Archaeology* 20(3), 589–600.
MacCannell, D. (2011) *The Ethics of Sightseeing*. University of California Press, Los Angeles, California.
Molina, A. and Rajagopal (2023) People, planet, and profit: Crossing the triple bottom line. In: *Challenge-Based Learning, Research, and Innovation*. Palgrave Studies of Entrepreneurship and Social Challenges in Developing Economies. Palgrave Macmillan, Cham, Switzerland, pp. 35–65.
Nurse, K. (2006) Culture as the fourth pillar of sustainable development. *Small States: Economic Review and Basic Statistics* 11, 28–40.
Orekat, F. (2015) Understanding visitors' experiences at religious heritage sites: A case study from Jordan. PhD thesis, The University of Sydney, Sydney, Australia. Available at: https://ses.library.usyd.edu.au/handle/2123/14718 (accessed 11 October 2024).
Packer, J. and Ballantyne, R. (2010) Conceptualizing the visitor experience: A review of literature and development of a multifaceted model. *Visitor Studies* 19(2), 128–143.
Pine II, B.J. and Gilmore, J.H. (1999) *The Experience Economy: Work is Theatre and Every Business a Stage*. Harvard Business School Press, Boston, Massachusetts.
Ramzy, M. and Morsi, A. (2020) Envisioning urban tourism sustainability in cultural heritage destinations. *Journal of Art & Architecture Research Studies* 1(1), 48–64.
Relph, E. (1976) *Place and Placelessness*. Pion Publishers, London.
Schmitt, B.H. (1999) *Experiential Marketing: How to Get Customers to Sense, Feel, Think, Act, Relate to Your Company and Brands*. The Free Press, New York.
Selstad, L. (2007) The social anthropology of the tourist experience. Exploring the 'middle role. *Scandinavian Journal of Hospitality and Tourism* 7(1), 19–33.
Sharkawi, A. and El-Rahman, M. (2022) Implementing the five senses as an innovation solution to enhance visitor experience in heritage sites case study El-Muizz street. *International Journal of Tourism, Archaeology and Hospitality* 2(3), 161–172.
Singer, D., Aaker, J. and Smith, A. (2011) The power of storytelling: What nonprofits can teach the private sector about social media. *McKinsey Quarterly* 1, 108–110.
Sotiriadou, P., Brouwers, J. and Le, T.A. (2014) Choosing a qualitative data analysis tool: A comparison of NVivo and Leximancer. *Annals of Leisure Research* 17(2), 218–234.

Strauss, A. and Corbin, J. (1998) *Basics of Qualitative Research Techniques*, 2nd edn. SAGE Publications, Inc, Thousand Oaks, California.
Swarbrook, J. and Horner, S. (2011) *Consumer Behaviour in Tourism*. Routledge, London.
Viljoen, A.H. and Kruger, M. (2020) Could conservation management be prioritised during captive wildlife experiences? *Acta Commercii* 20(1), a853.
Volo, S. (2009) Conceptualizing experience: A tourist based approach. *Journal of Hospitality Marketing & Management* 18(2–3), 111–126.
Walls, A.R., Okumus, F., Wang, Y. (Raymond) and Kwun, D.J.-W. (2011) An epistemological view of consumer experiences. *International Journal of Hospitality Management* 30(1), 10–21.

3 Motorcycle Tourism and Sustainability

Werner Gronau[1]*, Jana Große Hokamp[2] and Knut Scherhag[3]
[1]*School of Business Studies, University of Applied Sciences Stralsund, Stralsund, Germany;* [2]*Kohl und Partner, Schwäbisch Gmünd, Germany;* [3]*Department of Tourism/Travel Management, University of Applied Sciences Worms, Worms, Germany*

Abstract
As the tourism sector is among the major global business sectors, the sustainability challenge must be addressed, especially when it comes to the ecological perspective. The constantly growing rate of carbon dioxide emissions is heavily driven by transport-related emissions strongly associated with the global tourism sector. However, it is especially 'drive tourism' – which includes motorcycle tourism – where social and economic aspects play an important role as well. Motorcycle tourism takes place predominantly in rural areas, on less frequented routes, where it contributes to positive economic effects for local tourism service providers. Cultural differences among motorcycle travellers make it difficult to market corresponding travel products uniformly and necessitate a closer look at the source markets. The target groups are very specific and differ fundamentally between leisure riders, who often only use their motorcycles for short trips at the weekend near their home, and vacation travellers. The latter group is looking for experiences during their tour, often offered in the visited regions as special interest products from various local service providers. Their high purchasing power and extensive travel experience makes them a promising, yet rather challenging customer group, especially in less frequently travelled periphery regions. Building on these aspects, this chapter advances understanding of the possible trajectories motorcycle tourism can take towards a more sustainable future.

Keywords: Drive tourism, niche tourism, experience, motorcycle, sustainability

Introduction

The sustainability debate today looks back on long historic development. Starting from the Rio Declaration on Environment and Development in 1992, the three major dimensions of sustainability have been outlined as economic, social and environmental. Right from the beginning it became obvious that these three pillars of sustainability cause certain conflicts as noted, for example, by Jischa (1998, p. 118) who stated: 'Without a doubt, environmental thinking often conflicts with economic considerations. (...) It is becoming clearer that the social question will also take a central role in political considerations. (...) The obvious conflicts among the three pillars of sustainability must constantly be reconsidered by society'. We currently see the conflict between the needs of residents and mass tourism in places like Tenerife, where residents are protesting against excessive tourism – but not against tourism in general. Tenerife has 2.2 million inhabitants and around 14 million visitors whereas tourism accounts for around 35%

*Corresponding author: werner.gronau@hochschule-stralsund.de

 Niche Tourism and Sustainability: Perspectives, Practices and Prospects (eds A. Farmaki *et al.*)
DOI: 10.1079/9781800626669.0003

of the region's gross domestic product (Jones, 2024). The protesters therefore want limits on tourist numbers and curbs on what they describe as uncontrolled development (e.g. new hotels, traffic jams) harmful for the environment and residents.

Despite the existing conflicts within the concept of sustainability, the tourism sector was among the first ones to adopt sustainability and proclaim the era of sustainable tourism. Since the mid-1990s, sustainable tourism has become a buzzword not just in development studies, but also in tourism research. Despite the broad range of literature on the topic, a decent theoretical framework of the so-called sustainable tourism has until today not been established. Instead of addressing the identified theoretical divide between the concepts of development, sustainability and tourism, exemplarily presented by Sharpley (2000), the scientific community preferred to do 'case studies which explore the ways of applying sustainable principles to practice, often through small eco- or alternative tourism projects, [that] provide at best a micro solution to what is essentially a macro problem' (Wheeller, 1991, p. 93). This approach resulted in the long term in the currently often practised approach to distinguish between an unsustainable mass tourism system and more sustainable niche markets such as ecotourism, geotourism, etc. The ongoing friction between overall society goals with regard to sustainability and the actual attitude and actions of the individual tourist towards this approach is still visible in everyday demand.

Researchers claim, ever since Moore and Cater's (1993) first statements about an increasing environmental consciousness of tourists, that there is a substantial interest among holidaymakers towards sustainable tourism products. Also, 30 years later in the context of the restart of the tourism industry after COVID-19, one could witness a debate on the era of sustainable tourism to come. Unfortunately, we still lack fully convincing evidence of a major shift in consumer attitudes towards the willingness to pay for more sustainable tourism products. The start of the new millennium refocused the debate, by outlining the relevance of voluntary commitment by suppliers in the form of various corporate social responsibility (CSR) approaches and the implementation of sustainable supply chain management (SSCM) in tourism. But based upon the study of Font and Cochrane (2005), it became obvious that the main challenge is to apply existing SSCM practices more widely in tourism. Quality is a key aspect of supplier selection by tour operators, but even though many sustainability issues have significant effects on the quality of holidays experienced by tour operators' clients, the tourism sector only rarely includes sustainability issues as part of the quality equation. At the same time, the specific conditions of tourism seem not to favour such approaches with Zhang *et al.* (2009) pointing out that 'the characteristics of tourism such as: coordination intensive, perishable, information intensive, product complexity, demand uncertainty and complex dynamics interferes quite a lot with SCM-issues'. In need of an alternative, research refocused on the demand side to foster a more sustainable tourism development. But research has demonstrated that tourists tend to be unwilling to adopt sustainable consumption behaviours and nor do they believe it their responsibility to do so (Miller *et al.*, 2010), with even those who typically adopt (or claim to adopt) green consumerist behaviour at home not doing so while on holiday (Juvan and Dolnicar, 2014).

Concluding this brief introduction on the challenges in the field of so-called sustainability in the tourism context, one must consider that:

- There is a clear conflict between the three dimensions of sustainability; a balanced development considering all three dimensions seems somehow unrealistic. Therefore, one might have to consider a prioritization of one category over another.
- There are attempts among suppliers, for example through CSR or SSCM, to improve their tourism products towards being more sustainable; nevertheless, the attempts are limited and implementation is challenging.
- There is a growing awareness among consumers in regard to sustainability – especially the ecological perspective, but the tourism context seems to be among the most challenging fields of adaptation.

Accepting the above-mentioned aspects means that an evaluation of sustainability in a specific tourism field must consider: a possible prioritization of specific dimensions; the sustainability of the supply side in the form of the product itself and the way it is produced; and, last but not least, the attitude and awareness of the specific target group. The basic willingness of travellers to opt for sustainable travel and destinations is currently not very high. According to a survey of German travellers, sustainability is a relevant criterion that significantly influences travel decisions for only 4% of travellers in 2022. For around 7% of the remaining travellers, sustainability does not currently play a role in their vacation trip (Kuhn *et al.*, 2023, p. 57). This chapter, therefore, addresses the preconditions of the specific field of motorcycle tourism in regard to all three pillars of sustainability, while starting off from the attitudes and motivations of the specific target group and the resulting demands for destinations as well as tourism providers. Building upon these relevant aspects and their probability of implementation, the chapter contributes understanding of the possible trajectories towards a more sustainable future of the sector.

Motorcycle Tourism Considering Sustainability

Motorcycle tourism as a research field has come a long way: from the early beginning of academic contributions when motorcycling was perceived as a 'gangs-thing', as a field of ethnographic studies, to the days when motorcycling was accepted as a general leisure activity and consequently became a part of drive tourism research (Gronau and Große Hokamp, 2022, p. 249). The focus of motorcycling as a drive tourism form also contributed to a growing interest in the field of business and tourism studies, as it defined a new paradigm for the field which still prevails today. As Carson and Schmallegger (2011) said: 'The major change in the global self-drive markets is a shift from the car as a form of least-cost transport towards the use of a variety of self-driven vehicles that add value to the tourism experience' (p. 364). In drive tourism, motorhomes, convertibles, classic cars and motorcycles are included. What these user groups have in common is that 'the journey is the destination' and that they primarily experience the journey with their vehicles, even if the factors influencing the motivation to travel vary from target group to target group (Scherhag, 2023, p. 85). The volume of leisure traffic is comparatively high, which means that the environmental impact in the destination areas is also high. For example, a Swiss study showed that around 50% of journey times and around 44% of kilometres travelled are leisure-related (Egger and Stückelberger, 2022, p. 148).

In many industrialized countries, motorcycling has today achieved the status of a leisure activity. In the case of riding a motorcycle, the journey is often the destination and arrival at a specific destination is often only of secondary importance. Therefore, there tends to be a high number of kilometres per day as well as throughout the journey. Considering the high number of kilometres per day, the ongoing growth of motorcycle tourism in general and the increasing number of specialized tour operators in motorbike tours, one might identify an increasing conflict between sustainability and motorcycle tourism. This growing conflict is also visible when analysing academic literature in the field, as outlined by Gronau and Große Hokamp (2022). The three fields of publications on motorcyclists have been identified, namely: (i) their relevance as a specific consumer group addressing various issues such as expectations, demographics or the role of communities; (ii) their potential to foster destination development; and (iii) the most recent one, motorcyclists as a threat resulting from the constantly growing numbers of motorcyclists – in other words, the negative impacts of motorcycle tourism, such as noise pollution and the resulting consequences in road closures. With respect to the above-mentioned effects on winding roads, hilly or mountainous regions, those areas providing curvy roads get increasingly in the focus. Especially the European Alps see an increasing pressure from motorcycle tourists. Initial studies such as Scuttari *et al.* (2019) and Lechner *et al.* (2020) have analysed the effects of motorcycle noise and the resulting annoyance in the Alps. Despite the positive economic effect of motorcycle tourism, there is a growing resistance resulting, for example, in a ban on motorcycles louder than 95 dB(A) on selected roads in the Austrian region of Tyrol while the Italian province of South Tyrol has even imposed general restrictions on panoramic Alpine roads during the summers of 2017 and 2018 (Scuttari *et al.*, 2021). As more and more motorcycle tour operators start using terms such as holiday and adventure for their marketing (Groß and Stengel, 2010, p. 210), a growing number of products even include 'an off-road experience' within their activities. Keeping in mind the studies on the behavioural response of wildlife to motorcycle traffic by Buxton *et al.* (2020), one might consider motorcycle tourism as a growing threat for tourism destinations.

Motorcycle Tourism Demand Perspectives

This section provides a deeper understanding on the attitudes and behaviours of different motorcycle tourists to evaluate whether the growing conflict in the field of environmental sustainability refers to the social and economic dimension as well. Furthermore, a better understanding of consumer behaviour in the field of motorcycle tourism might also help to overcome some of the challenges, while also providing evidence for the effectiveness of certain measures to counter negative developments.

Aspects of motorcycle tourism demand impacting destination management

Before profiling the specific aspects when it comes to motorcycle tourism demand, there first follows a short wrap-up of the conceptual context of demand. Demand is based on the needs of the customer (Dann, 1977; Uysal and Hagan, 1993; Kassean and Gassita, 2013), which can be defined as 'a state of deprivation in a person' (Kotler *et al.*, 2017, p. 31). This can be a physical basic need as well as psychological (Holloway and Humphreys, 2016). Demand results from the push factors, which represent the socio-psychological state of the individual (Dann, 1977; Uysal and Hagan, 1993; Yoon and Uysal, 2005; Kassean and Gassita, 2013). In academic literature, there are many ways of measuring demand for tourism products, but number of visitors and spending within the destination are dominant variables (Yoon and Uysal, 2005). The total demand consists of many different types of people, with different preferences, tastes and spending power (Doswell, 1997). To attract visitors, their motivations need to be understood, and a strong focus needs to be set on the emotions of the visitor and their satisfaction (Yoon and Uysal, 2005). Push factors are the driving forces that compel the customer to satisfy a need. Needs that are satisfied through travelling are very individual (Kassean and Gassita, 2013). Not one destination is able to satisfy the needs of the vast amount of different people and different forms of tourism (Ferreira, 2015). Therefore, it is necessary for service providers in a destination to segment the market in the sense of providing special interest (niche) products, because only a few destination management organizations (DMOs) are able to specifically address several target groups. Doswell (1997, p. 36) presents three reasons to segment: (i) 'to develop the product to satisfy specific needs'; (ii) 'to identify where and how to sell the product'; and (iii) 'to judge if and how to mix different segments together'. Crompton and McKay (1997) state the same reasons with the slight difference that the third reason is understanding the decision process of the customer.

Which criteria are used for segmenting can be individually decided, as long as it makes sense, to identify the behaviour and needs (Doswell, 1997). Each group that has been segmented and defined as a target group needs to be analysed and understood (Kassean and Gassita, 2013). It is vital for the stakeholders in a destination to have the knowledge about travel behaviour and motivations (Holyoak and Carson, 2009). Motivations of consumers are the subject of research for many academic writers. There is general agreement that the needs and wants of tourists influence the motivation to travel and tourism behaviour and is a key priority in analysing the segments (Doswell, 1997; Kassean and Gassita, 2013). However, more information is required for full analysis of motivations according to Yoon and Uysal (2005). These motivations are not rigid but can change over time. These changes develop due to economic, demographic and environmental change or due to different product offers by competing regions or stakeholders (Tsiotsou and Goldsmith, 2012; Bieger and Beritelli, 2013), just as they also affect the destination. There is the need for a destination to consider the analysis of the segmented groups again over time (Tsiotsou and Goldsmith, 2012). If the customer needs are satisfied by a destination, they will remember and share the experience with friends and family, or will return to the destination (Oppermann, 2000). Word of mouth, the recommendation of the destination, is one of the most reliable and most often sought types of information for potential customers (Yoon and Uysal, 2005). Whether the tourist is of the opinion that the destination can satisfy their needs or not, is the determining factor influencing the decision of travelling to the destination or not.

Demographics of motorcycle tourism demand

Different academic studies on motorcycle tourists found that the motorcycle tourist is often above the age of 40 years (Frash *et al.*, 2018; Scherhag *et al.*, 2018; Merry *et al.*, 2020). Broughton and Walker (2009) found their peak at the age 30–39 years with a quarter above the age of 50, while Weddell (2014) reported an average age of 56 years. These studies also state that the group is growing older. This entails that motorcyclists have more leisure time and more disposable income to spend (Cater, 2012, 2017b) because family responsibilities have decreased again after the children are out of the house and care for themselves. Way *et al.* (2010) state that the age group of 50–60 years is the fastest-growing group within motorcycle tourism. While some riders have been riding since teenage years, many have picked up/restarted motorcycle riding in recent years to fulfil lifelong dreams or fantasies that family commitments and responsibilities had put on hold (Way *et al.*, 2010). This change in motivation functions as a push factor to take the motorcycle and fulfil a lifelong dream that is not limited to the hometown but extends to other destinations.

The sector is still rather associated with and dominated by men, but women are seen as a growing market (Pinch and Reimer, 2012). There are more women joining the group of motorcyclists in recent years (Cater, 2012, 2017a; Sykes and Kelly, 2012), often at a later age. Reasons for the smaller number of female motorcyclists could be not being confident in handling a motorcycle or fearing not being socially accepted (Broughton and Walker, 2009) but also due to family commitments. According to Sykes and Kelly (2016), the motorcyclist in the USA spends generally between US$101 and US$250 per day, and only 5% of motorcyclists spend more than US$250. This is in line with Cater (2012, 2017b), studies in the UK which indicate that £99 (roughly US$128) was spent daily including accommodation. Weddell (2014) found that for an entire trip of roughly four nights, more than US$1000 was spent. Broughton and Walker (2009) state that the normal motorcycle tourist earns a little above average in the UK. This excludes the over-represented group of retired persons who worked in upper and managerial positions. Most motorcyclists, according to Weddell (2014), had finished some kind of college degree. This can explain the higher disposable income that the motorcycle tourist has available. In a comparative study, Scherhag and Frash (2017) showed that the household income of motorbike travellers in Germany is around 80% – and, in the USA, around 74% – higher than the average household income in the respective countries. Most of the motorcyclists spent on average over £4000 per year for their motorcycle and motorcycle-related activities.

The demographics described here are confirmed in other studies about motorcycle tourists that do not necessarily focus on motorcycle tourism (e.g. Groß and Stengel, 2010). Changes based on shifts in demographics, economic and environmental factors or through the rise of other products also affect the target group (Tsiotsou and Goldsmith, 2012; Bieger and Beritelli, 2013). Upon growing older the needs of the customer change and with that so do the push and pull factors. It is believed that ageing will result in the need for comfort rising, which means that the motorcyclist today may not be able or may not want to drive as much in the future. This can result in a decrease of the radius that is driven by them. The demand for more alternative attractions fitting for them in the destination is rising as a demand for luxury products such as wellness, better accommodation and food. In summary, the motorcycle tourist group is mainly made up of individuals above the age of 40 years with a higher number of men, who earn a little above average. They can afford the motorcycle and are willing to spend money on motorcycle activities. Growing older is a natural development but with the motorcycle tourist there is the challenge that there is only a small number of replacements by the younger generation.

Motivation and goals of motorcycle tourism demand

Different studies have confirmed that, in motorcycle tourism, the focus is on riding for pleasure (Broughton and Stradling, 2005; Walker, 2011; Sykes and Kelly, 2016; Scol, 2016; Cater, 2017a). Instead of using the motorcycle solely as an individual means of transportation, the experience of the journey in itself is the main goal (Walker, 2011). However, the goals and fulfilling needs that represent the push factors for motorcycle holidays differ among groups of motorcyclists (Broughton and Stradling, 2005). Push factors can include escaping the daily routine (Frash *et al.*, 2018) by travelling to different places. This contrasts with other tourists who do not want to lose time travelling. Motorcycle tourists find enjoyment in riding and their goal is consuming, challenging and defeating the road as a physical object (Scol, 2016). Additionally, there are motorcyclists who just go out to take a ride in the summer, without having a specific destination. For them the goal is also to enjoy the ride and return home safely (Broughton and Stradling, 2005). For the motorcycle rider it is a way of escaping from the methodological daily routine (Frash *et al.*, 2018). It is this leisure time and the passion, which are rated higher than the utilization of the motorcycle (Scherhag *et al.*, 2018), that lead to the overall goal of happiness in the end (Broughton and Walker, 2009). The differences in the motivation and needs are to certain extent already recognized when buying a specific motorcycle. There are perceived differences in the brand that the rider chooses and the different behaviour of riders of different brands.

Based on age and gender, differences in enjoyment within the motorcycle group can be found. Austin (2009) mentions differences in connection to motorcycle events. Broughton and Walker (2009) distinguish between challenge-based enjoyment, where mastering the curvy roads, bends and scenery are more important, and rush-based enjoyment, where the speed is important. The motorcyclist who is looking for challenge-based enjoyment will not necessarily buy a sports machine with high speed which would be the choice of the rush-based enjoyment seeker. In general, more motorcyclists find their enjoyment in challenge-based situations with the curvy roads but also the elevation and the landscape, which is often very attractive for motorcyclists in mountainous areas (Broughton and Walker, 2009; Scol, 2016; Scherhag *et al.*, 2018). In addition, the lifestyle and attractiveness of events can be a major motivation for motorcyclists (Sykes and Kelly, 2014; Weddell, 2014). This translates directly into push factors of motorcycle tourists. Attractive destinations are areas where these driving needs can be satisfied, namely mountainous areas with many scenic, curvy roads. Additionally, lifestyle and events are push factors for many motorcyclists. There is also a difference between the home-based motorcycle tourist and the tour driver which also makes a difference in the push and pull factors. Home-based motorcycle tourists have a stronger focus on their destination and on the facilities on site; they are mostly riding on their home track and visit the well-known motorcycle meeting places. The tourer who will just stop by for one or two nights during their holiday trip/short trip often looks for different places because they are more interested in cultural and natural highlights and will see something unexpected.

The role of community in motorcycle tourism demand

Cater (2017b) found that in Wales over 40% of motorcyclists are members in a motorcycle organization such as a local motorcycle club, a national or motorcycle-specific club or action groups. This membership and being part of the motorcycle community is described as positive by many motorcyclists (Broughton and Walker, 2009; Cater, 2017b). There are many different motorcycle clubs and communities, which have to be distinguished from the gangs that have a rather negative image (Coleman, 2010). A strong cohesion within the community can be found between the motorcyclists, with orientation towards social factors (Coleman, 2010; Cater, 2017b), solidarity and collective identity. These communities are non-geographic (Austin and Gagné, 2007) and share stories, experiences, rituals and markets for clothing (Haigh and Crowther, 2005; Austin, 2009). There seems to be no segmentation based on the motorcycle type or brand, as the motorcyclists have generally more in common than they have disparities (Walker, 2011).

The strong social cohesion of the motorcycle community was highly visible in 2020. In the district of Reutte in Austria, the Tyrol government decided to close six routes to motorcycles with a standing noise above 95 dB(A) on short notice based on a sound study that was conducted in the area in 2018 and 2019 (Lechner *et al.*, 2020). This regulation increased the long discussion of sound pollution by motorcycles in Europe again. In Germany, demonstrations took place throughout the whole country against possible regulations that were discussed for different areas. On the one hand, there are a number of motorbikes that have a higher stationary noise level at the factory but are perfectly legal to drive, as there are no limit values for stationary noise in Europe, in contrast to driving noise. On the other hand, there are several cars with higher stationary noise level (>95 dB(A)) but are not subject to a driving ban (Scherhag, 2023). At this point, the Swiss National Council is one step ahead, as it wants to reduce unnecessary engine noise regardless of the polluter, but also wants to penalize the emitters (Baumann, 2021). 'Too many' motorbikes can be justified with the argument of an increase in the frequency of accidents and the associated overloading of the emergency services, as well as the need to reduce carbon dioxide emissions. A similar argument is used in the Italian province of South Tyrol, where the focus appears to be on the possibility of generating revenue (tolls). The aim there is to impose quotas on vehicles from 2024 and exclude particularly noisy vehicles and to set and adhere to a maximum volume of traffic by means of 'bookable roads' (Kratschmar, 2023). While in Tyrol, tourism stakeholders and regions had many cancellations from the motorcycle community. Despite everything, the designated roads will be closed again in 2024 (Kratschmar, 2024).

This community aspect is also visible in motorcyclists' travel behaviour. Most motorcyclists travel in groups (Cater, 2017a, b) that are either planned or met at the destination. The group sizes can differ greatly. While Cater (2017a, b) states an average group size of up to five individuals with approximately a quarter of the groups being bigger, Weddell (2014) states an average of four individuals. Differences arise due to different research locations, as Scherhag *et al.* (2018) found in comparing Germany and the USA. The average group size is similar to the studies mentioned before but in Germany 83% of motorcyclists travel in groups while in the USA only 39% travel in groups. Another difference is visible when looking at the portion of motorcyclists carrying a pillion rider. The USA has a higher percentage of 39% compared to 22% in Germany. Weddell (2014) stated that nearly 60% travel with a partner either as a pillion rider or as a second driver. Also, in Cater's (2017a, b) study roughly half of the individuals carry a pillion rider half of the time. This has an influence on the size of the group but also on the demand, as the pillion rider also has needs that can differ from those of the driver. In addition, the cultural differences of motorcycle tourists from different countries/regions of origin must be considered for market development in order to address the target groups as efficiently as possible (Scherhag *et al.*, 2022).

In motorcycle tourism, word of mouth plays a key role in decision making. It is seen as one of the most reliable and most sought-after information sources (Yoon and Uysal, 2005) when planning a trip (Frash *et al.*, 2018). Cater (2017b) found in his study that motorcycle tourists trust the information they receive from other motorcyclists. Additionally, they focus their research on websites that fit their lifestyle. Weddell's (2014) study concluded that the main source of information is the Internet followed by previous visits and then friends. In total, this shows that the main marketing tools for motorcycle tourists are word of mouth and the Internet. Even though motorcycle riding is a solo activity, there is the strong community belonging which already satisfies a need of the rider (Broughton and Walker, 2009). The trips taken can then also be done in groups which gives the rider a specific identity (Walker, 2011). Within the clubs and communities, which take on different forms (Walker, 2011), stories about different trips, experiences and destinations will be shared (Scol, 2016; Cater, 2017a). The close-knit nature of the community also becomes clear in Cater's (2017b) study, where the motorcyclists stressed that recommendations by other motorcyclists were often considered and used by them. The importance of word of mouth and the preferences in information flows influence the presentation of supply at destinations that is an integral part of the pull factors.

Requirements towards the journey in motorcycle tourism demand

Motorcycle tourists have special needs and requirements towards the destination as well as towards the journey itself, which differ from other tourist target groups (Sykes and Kelly, 2014). Knowing these requirements specifically for the motorcyclists can influence the presentation of the supply by a destination. These illustrations are the strength of the pull factors. In current academic research, there is a great deal of expectation by motorcyclists towards natural circumstances and the original offer by a destination. The choice for a destination is based on these expectations and the presentation of supply (Yoon and Uysal, 2005). In Weddell's (2014) study, motorcyclists stated that when travelling in mountain areas, the scenery and views, being in nature and getting away from traffic, as well as challenging roads and visiting touristic sights or scenic stopping points, are important to them (Cater, 2012; Weddell, 2014; Sykes and Kelly, 2016; Scol, 2016).

The motorcycle tourist travels in a greater radius than tourists in many other tourism segments, as previously mentioned. This is especially the case with the motorcycle tourists who are home-based riders (they book a central hotel and organize excursions/day trips in the destination from there), as they stay several days in the region and travel around. Deduced from this are pull factors for home-based motorcycle tourists. These pull factors include, additionally to the natural/environmental factors, a large variety of different tours and attractions. In the academic literature, there is no clear information on how large the radius is around the destination that is reachable by the motorcycle tourist. The High-Bike Testcenter has collected the kilometres that their customers covered for 1 day. Based on this data between 2011 and 2020, the average has been calculated to receive an average radius per day. For this motorcycle target group, the average kilometres that are driven per day is 315.8 km. This results in a radius of approximately 158 km around the accommodation as home base. Within this radius many other attractions, like coffee shops or best viewing spots, will be taken into consideration when planning the tour (Große Hokamp, 2021).

A quality landscape with scenic and beautiful views highlights the pleasure of riding. Cater (2017b) names the curvy roads as the main goal and therefore the main pull factor of the destination as the preferred route of the motorcyclists. Preferable are also good road conditions and elevation changes with signages and local attractions (Schouten and McAlexander, 1995; Austin, 2009; Walker, 2011; Weddell, 2014; Sykes and Kelly, 2014, 2016; Cater, 2017b). Beautiful landscapes are also important during day trips in a destination (Scherhag *et al.*, 2022). The motorcyclist wants to feel welcome in the place (Sykes and Kelly, 2014; Weddell, 2014; Scol, 2016; Cater, 2017a, b).

The facilities that the motorcyclists are looking for during their tour can differ according to the type of traveller, as mentioned already. There are different push factors that are very specific and individual for each type of motorcycle traveller. An analysis of these is outside the scope of this work. But a differentiation in demand can be made to a certain extent. Motorcyclists are looking for authentic experiences and accommodation that include family-owned accommodation rather than hotel chains. Cater (2017b) as well as Weddell (2014) identified that most trips motorcyclists take are day trips and overnight trips up to four to five nights away either from their home country or overseas (Scherhag and Frash, 2017). When selecting accommodation, the proximity to other facilities such as restaurants and pubs is wanted (Weddell, 2014). Important facilities also include appropriate and safe parking spots, preferably with a covered parking space for motorcycles (Austin, 2009; Walker, 2011; Sykes and Kelly, 2014; Weddell, 2014; Cater, 2017b; Cappelloni *et al.*, 2019). Additionally, a cleaning station with towels and other tools is also attractive to motorcycle tourists. Motorcycle-friendly accommodation is preferred because they focus especially on the target group and therefore often have preferred facilities (Weddell, 2014).

In any case, motorcyclists are a social group of tourists and are therefore longing to meet other motorcyclists. This is possible when arriving at the hotel from a tour, during meals or in the evening at the bar or in a dedicated space at the hotel. There is a very high chance that other motorcyclists meet to socialize, and the hosts are familiar with motorcycle riding, generally riding themselves (Sykes and Kelly, 2016; Frash *et al.*, 2018; Scherhag *et al.*, 2018). Hosts are a very trusting and a close source of information onsite for the motorcycle tourists. They know the region very well and can give insider tips as well as more specific information than any booklet or online platform. The discussions about the motorcycles and the tours, the so-called 'Benzingespräche' in German, are an important feature for motorcyclists during their holiday. It can be beneficial for a motorcycle-friendly accommodation to have a restaurant or another dedicated place where like-minded motorcycle tourists can meet (Scherhag *et al.*, 2018). For a region not only the attractions that are available in the surroundings can function as additional pull factors, but especially events can serve as a factor attracting motorcycle tourists (Holloway and Humphreys, 2016), especially since these are often the places and occasions where motorcyclists find like-minded people.

The role of image in motorcycle tourism demand

Motorcyclists are not always welcome at a destination. The image of motorcyclists, which has been strongly influenced by the media, is often perceived as negative (Broughton and Walker, 2009). This negative image developed from the behaviour of a minority of motorcyclists with antisocial behaviour (Broughton and Walker, 2009; Quadri-Felitti *et al.*, 2019) – often speeding, with manipulation of the exhaust system. Additional influence from movies such as *The Wild One* (1954) and *Easy Rider* (1969) portraying the motorcyclist in connection with delinquency, crime and social threats is to blame (Pinch and Reimer, 2012). Around 1% of motorcyclists exhibit this negative behaviour that shapes the perception of motorcycle tourists and influences business attitudes towards this target group. In media, this 1% is often presented while 99% of the motorcyclists are law-abiding (Barker, 2017). Additionally, residents along favoured motorcycle routes can develop a negative attitude towards motorcyclists due to sound pollution as in the case of 'Außerfern' in Tyrol (Lechner *et al.*, 2020). This also depends on the specific extent of the behaviour of the motorcyclists driving there. This image influences the marketing of motorcycle tourism and especially the atmosphere at the destination and has therefore a direct impact on pull factors. Managing these perceptions is an integral part of increasing pull factors for a destination.

It is needful that more information is communicated to motorcyclists, also regarding the health hazards of noise, which have proven to exist. There is a consensus among motorcyclists that it is only a small group of 'black sheep' that jeopardize the image of motorcyclists in a negative way. There are also calls from the elder motorcycle riders for higher penalties for tampering with motorcycles (Scherhag, 2023). A lot of DMOs react differently to the potential for conflict that the target group of motorcyclists brings to tourism regions due to non-compliant behaviour. For motorcycle travellers, unpolluted nature with attractive landscapes and winding roads with little traffic are important criteria when choosing a destination for a trip (Scherhag *et al.*, 2022; Scherhag, 2024). Despite the economic importance of the target group for accommodation and gastronomy businesses, there is no marketing for the motorcycle traveller target group by tourism organizations in many destinations. In other places, like South Tyrol, motorcycle tourism has evolved despite decisions to introduce charges (tolls) for the use of endangered Dolomite roads. Such measures, although contested, intend to reduce both noise and carbon dioxide pollution and improve the sustainability contribution of motorcycle tourism (Kratschmar, 2023).

Towards a More Sustainable Motorcycle Tourism

As elaborated earlier, motorcycle tourism is a tourism segment with significantly higher spending than many other tourism forms. In addition, the preferred regional settings are rural periphery regions with a low amount of traffic on their twisty roads, resulting in increased tourism spending in less travelled regions. In combination with the high affinity towards commercial accommodation, local restaurants and frequent breaks during their rides, motorcycle tourists have a significant economic impact within their specific regional context. This aspect is important to communicate to destination areas as well as local residents. This can help to reduce resistance to individual tourism segments, as it is recognized that without tourism or with fewer target groups, the economic situation in the destination areas will become more dire. The large number of niche tourism offers in particular are increasingly contributing to an expansion of target groups in a destination, as special service providers primarily address these via target-group-specific social media channels. These channels are also used to communicate holiday experiences. At this point, it should be noted that destinations are not just a holiday destination, but also the living space of people who want to feel at home there. Therefore, awareness has to be created among riders as well in order to overcome some of the downsides of the sector.

Campaigns from destinations towards the reduction of noise have been well received among a vast share of riders and already contributed to a reduction of the issue. When addressing noise, but also air pollution, one might interpret the increasing number of motorcyclists as an environmental threat, but the strong interest in riding their own motorcycle combined with the understanding of the journey as an integral part of the experience makes this group a rather reluctant one to the much more harmful air travel. The current debates on the noise pollution in several European countries have already triggered positive results in terms of initiatives among riders to create more awareness on the issue while also campaigning against the small minority of riders causing the problem. Also, first steps towards a legal framework fighting noise and air pollution of traffic in general are on their way, as evidenced for example in Switzerland. Such initiatives will help to reduce the negative impacts of individual motorized transportation in general, which is an important step when considering the still leading role of this mode of transportation when it comes to holidaymaking. Germany might work as a good example, where the private motor vehicle still accounts for around 46% of all holiday trips, followed by the airplane with 42% (FUR, 2023).

In general, a change towards greener forms of transportation in the tourism context is rather unlikely, as the car gets more and more replaced by the plane (Scherhag, 2024). Beyond the specific case of transportation options, one also has to admit that sustainability in general is a rather neglectable aspect when it comes to decision making among holidaymakers. Once more, Germany might work as a typical example for the overall Western European travel market wherein only 4% of all travellers in 2022 considered sustainability as a decisive factor for their holiday trips of 5 days or more, in contrast to 61% of the population who have a generally positive attitude towards sustainability when travelling on holiday (Kuhn *et al.*, 2023). Therefore, one has to admit that the already existing activities and campaigns in motorcycle tourism towards reducing the negative impacts of this transportation mode, as well as the fact that this segment will not see a major shift towards even more emission-intensive forms of transportation, are positive. Putting the debate on more sustainable forms of transportation in the broader context, one might also refer to the already existing alternatives to fossil-fuel-powered vehicles. At present, electric motorbikes are still far too expensive for the mass market and also still lack range, but when considering the rapid developments in the car segment, one might already anticipate a major change in powering motorcycles by the end of this century.

Beyond the economic and the environmental dimensions, motorcycle tourism has a social dimension as well. As described earlier, motorcyclists can in general be described as social animals. Even though motorcycle riding is a solo activity, they like to travel in groups, mingle with locals and want to experience the uniqueness of the region they are in. As motorcycle tourism strongly relies on existing road and tourism infrastructure such as accommodation or food and beverage outlets, its significance socially differs only minorly from other forms of tourism. Nevertheless, one might argue that due to the main focus on rural and periphery regions this niche tourism form might contribute to a more equal allocation of tourism expenditures. While favouring the locally owned and operated small and middle-sized tourism business (which are highly common in rural and peripheral regions), there might be a more positive effect on local employment and income than in large-scale international providers common in tourism hotspots.

References

Austin, D.M. (2009) Ritual and boundary distinction in a recreational community: A case study of motorcycle rallies and riders. *Qualitative Sociology Review* 5(2), 70–93.

Austin, D.M. and Gagné, P. (2007) Community in a mobile subculture: The world of the touring motorcyclist. *Studies in Symbolic Interaction* 30, 411–437.

Barker, T. (2017) Motorcycle clubs or criminal gangs on wheels? In: Bain, A. and Lauchs, M. (eds) *Understanding the Outlaw Motorcycle Gangs*. Carolina Academic Press, Durham, North Carolina, pp. 7–28.

Baumann, U. (2021) Schweiz gegen laute Motorräder. Nationalrat forciert Kampf gegen Motorenlärm. *Motorrad-Online,* 28 June. Available at: https://www.motorradonline.de/ratgeber/gegen-laute-motorraeder-initiative-schweiz/ (accessed 28 July 2021).

Bieger, T. and Beritelli, P. (2013) *Management von Destinationen*, 8th edn. Oldenbourg, Munich, Germany.

Broughton, P. and Walker, L. (2009) *Motorcycling and Leisure: Understanding the Recreational PTW Rider*. CRC Press, Boca Raton, Florida.

Broughton, P.S. and Stradling, S. (2005) Why ride powered two wheelers. In: *Behavioural Research in Road Safety, Fifteenth Seminar*. Department for Transport, London, pp. 68–78.

Buxton, R.T., McKenna, M.F., Brown, E., Ohms, R., Hammesfahr, A. *et al*. (2020) Varying behavioral responses of wildlife to motorcycle traffic. *Global Ecology and Conservation* 21, e00844.

Cappelloni, F., Staffieri, S. and Baglioni, F. (2019) Knowledge of mototourist's habits to develop a successful tourist destination: An Italian case study. In: Ekinci, Y., Sharples, L., Viglia, G. and Gursoy, D. (eds) *9th Advances in Hospitality and Tourism Marketing and Management Conference Proceedings*, University of Portsmouth, Portsmouth, UK, pp. 250–261.

Carson, D. and Schmallegger, D. (2011) Drive tourism: A view from the road. In: Carson, D. and Prideaux, B. (eds) *Drive Tourism: Trends and Emerging Markets*. Routledge, London, pp. 358–368.

Cater, C. (2012) *Motorcycle Tourism in Ceredigion*. Tourism Society Wales, Cardiff.

Cater, C.I. (2017a) Long way up: Powered two-wheeled journeys in northern peripheries. In: Weaver, D., Prebensen, N.K. and Lee, Y.-S. (eds) *Arctic Tourism Experiences: Production, Consumption and Sustainability*. CAB International, Wallingford, UK, pp. 149–158.

Cater, C.I. (2017b) Tourism on two wheels: Patterns of motorcycle leisure in Wales. *Tourism Management* 61, 180–189.

Coleman, J. (2010) Bikers; Who are these people? Summer Research Paper 32.

Crompton, J.L. and McKay, S.L. (1997) Motives of visitors attending festival events. *Annals of Tourism Research* 24(2), 425–439.

Dann, G.M.S. (1977) Anomie, ego-enhancement and tourism. *Annals of Tourism Research* 4(4), 184–194.

Doswell, R. (1997) *Tourism: How Effective Management Makes the Difference*. Butterworth-Heinemann, Oxford.

Egger, T. and Stückelberger, U. (2022) Der touristische Verkehr – Vom Stiefkind zum Trumpffaktor. In: Bieger, T., Beritelli, P. and Laesser, C. (eds) *Krisenbewältigung und Digitale Innovationen im Alpinen Tourismus*. ESV, Berlin, pp. 147–158.

Ferreira, D.P. (2015) Push- and pull forces within outbound destination choice. Master's thesis, Nelson Mandela Metropolitan University, Port Elizabeth, South Africa.

Font, X. and Cochrane, J. (2005) *Integrating Sustainability into Business: A Management Guide for Tour Operations*. United Nations Environment Programme and Tour Operators Initiative for Sustainable Tourism Development, Paris.

Frash, R.E., Blose, J.E., Smith, W.W. and Scherhag, K. (2018) A multidisciplinary marketing profile of motorcycle tourists: Explorers escaping routine to find flow on scenic routes. *Tourism Recreation Research* 43(4), 432–444.

FUR (Forschungsgemeinschaft Urlaub und Reisen e.V.) (2023) Reiseanalyse 2023 - Selected first results of the 53rd Reiseanalyse. Available at: https://reiseanalyse.de/wp-content/uploads/2023/03/RA2023_Erste_Ergebnisse_Broschuere_EN-1.pdf (accessed 22 December 2023).

Gronau, W. and Große Hokamp, J. (2022) Motorcycle tourism: The long ride towards an academic field of research – a literature review. *Zeitschrift für Tourismuswissenschaft* 14(3), 248–262.

Groß, S. and Stengel, N. (2010) *Mietfahrzeuge im Tourismus – Grundlagen, Geschäftsprozesse und Marktanalyse*. De Gruyter/Oldenbourg, Munich, Germany.

Große Hokamp, J. (2021) Motorcycle tourism: Curse or blessing?. Master's thesis, University of Applied Sciences, Stralsund, Germany.

Haigh, J. and Crowther, G. (2005) Interpreting motorcycling through its embodiment in life story narratives. *Journal of Marketing Management* 21(5–6), 555–572.

Holloway, J.C. and Humphreys, C. (2016) *The Business of Tourism*, 10th edn. Pearson Education Limited, Harlow, UK.

Holyoak, N. and Carson, D. (2009) Modelling self-drive tourist patterns in desert Australia. In: *ATRF 2009, 32nd Australasian Transport Research Forum, 29 September–1 October 2009, Auckland, New Zealand*. Available at: https://australasiantransportresearchforum.org.au/wp-content/uploads/2022/03/2009_Holyoak_Carson.pdf (accessed 14 October 2024).

Jischa, M. (1998) Sustainable development: Environmental, economic and social aspects. *Global Journal of Engineering Education* 2, 115–124.

Jones, S. (2024) Tens of thousands protest against Canary Islands' 'unsustainable' tourism model. *The Guardian Online,* 24 April. Available at: https://www.theguardian.com/world/2024/apr/20/thousands-protest-canary-islands-unsustainable-tourism (accessed 14 October 2024).

Juvan, E. and Dolnicar, S. (2014) The attitude–behaviour gap in sustainable tourism. *Annals of Tourism Research* 48, 76–95.

Kassean, H. and Gassita, R. (2013) Exploring tourists' 'push and pull' motivations to visit Mauritius as a holiday destination. *Tourismos* 8, 39–56.

Kotler, P., Bowen, J. and Makens, J.C. (2017) *Marketing for Hospitality and Tourism*, 7th edn. Pearson Education Limited, Harlow, UK.

Kratschmar, J. (2023) Weniger Lärm in den Dolomiten. Fahrzeuge-Kontingentierung frühestens 2024. In: *Motorrad-Online,* 7 July. Available at: https://www.motorradonline.de/reise/laerm-dolomiten-low-emission-zone/ (accessed 7 July 2023).

Kratschmar, J. (2024) Fahrverbote über 95db(A) Standgeräusch in Tirol. Diese Bikes dürfen 2024 nach Tirol. In: *Motorrad-Online,* 5 January. Available at: https://www.motorradonline.de/ratgeber/motorrad-laerm-fahrverbot-oesterreich-tirol-modelle-2024/ (accessed 5 January 2024).

Kuhn, F., Günther, W., Braun, M., Yarar, N., Schmücker, D. *et al.* (2023) *Nachhaltigkeit bei Urlaubsreisen: Bewusstseins und Nachfrageentwicklung und Ihre Einflussfaktoren – Monitoringbericht auf Basis von Daten Der Reiseanalyse 2023*. Forschungsgemeinschaft Urlaub und Reisen e.V, Kiel, Germany.

Lechner, C., Schnaiter, D., Siebert, U. and Böse-O'Reilly, S. (2020) Effects of motorcycle noise on annoyance – a cross-sectional study in the Alps. *International Journal of Environmental Research and Public Health* 17(5), 1580.

Merry, K., Bettinger, P., Siry, J. and Bowker, J.M. (2020) Preferences of motorcyclists to views of managed, rural southern United States landscapes. *Journal of Outdoor Recreation and Tourism* 29, 100259.

Miller, G., Rathouse, K., Scarles, C., Holmes, K. and Tribe, J. (2010) Public understanding of sustainable tourism. *Annals of Tourism Research* 37(3), 627–645.

Moore, S. and Cater, B. (1993) Ecotourism in the 21st century. *Tourism Management* 14(2), 123–130.

Oppermann, M. (2000) Tourism destination loyalty. *Journal of Travel Research* 39(1), 78–84.

Pinch, P. and Reimer, S. (2012) Moto-mobilities: Geographies of the motorcycle and motorcyclists. *Mobilities* 7(3), 439–457.

Quadri-Felitti, D., Sykes, D. and Chen, F. (2019) Is motorcycle tourism ready to rev up in Pennsylvania? An exploratory study of suppliers' business attitudes of motorcycle tourism. *International Journal of Culture, Tourism and Hospitality Research* 13(1), 1–15.

Scherhag, K. (2023) Motorradlärm als konfliktfeld im regionalen tourismus. In: Bandi Tanner, M., Wirth, S. and Roller, M. (eds) *Tourismus im Wandel – Klimaschutz und Nachhaltige Entwicklung*. ESV, Berlin, pp. 83–97.

Scherhag, K. (2024) Motorcycle tourism – differentiation and classification. *Transportation Research Procedia* 80, 145–153.

Scherhag, K., Gross, S. and Sand, M. (2022) Adventures on two wheels – comparative study of motorcycle adventure tourists in Germany, Austria and Switzerland. *Zeitschrift für Tourismuswissenschaft* 14(3), 303–332.

Scherhag, K. and Frash, R.E. (2017) Motorcycle tourism – empirical study on travel motivation in Germany and the USA. In: *2017 ITB Berlin Convention, Berlin, Germany, 8–12 March 2017.*

Scherhag, K., Schröder, A. and Frash, R.E., Jr (2018) Anforderungen von Motorradtouristen an Beherbergungsbetriebe – Erste Ergebnisse einer deutsch-amerikanischen Vergleichsstudie. In: Ehlen, T. and Scherhag, K. (eds) *Aktuelle Herausforderungen in Der Hotellerie: Innovationen und Trends*. ESV, Berlin, pp. 249–260.

Schouten, J.W. and McAlexander, J.H. (1995) Subcultures of consumption: An ethnography of the new bikers. *Journal of Consumer Research* 22(1), 43–61.

Scol, J. (2016) Motorcycle tourism: Renewed geographies of a marginal tourism practice. *Via Tourism Review* 9. DOI: 10.4000/viatourism.356.

Scuttari, A., Orsi, F. and Bassani, R. (2019) Assessing the tourism–traffic paradox in mountain destinations: A stated preference survey on the Dolomites' passes (Italy). *Journal of Sustainable Tourism* 27(2), 241–257.

Scuttari, A., Pechlaner, H. and Erschbamer, G. (2021) Destination design: A heuristic case study approach to sustainability-oriented innovation. *Annals of Tourism Research* 86, 103068.

Sharpley, R. (2000) Tourism and sustainable development: Exploring the theoretical divide. *Journal of Sustainable Tourism* 8(1), 1–19.

Sykes, D. and Kelly, K. (2012) Mapping the study of motorcycle tourism: Impacts and opportunities. In: *Proceedings of the International Conference on Tourism and Events: Opportunities, Impacts and Change, Belfast, Northern Ireland, 20–22 June 2012*, University of Ulster, Belfast, UK, pp. 543–557.

Sykes, D. and Kelly, K.G. (2016) Motorcycle drive tourism leading to rural tourism opportunities. *Tourism Economics* 22(3), 543–557.

Sykes, D.M. and Kelly, K.G. (2014) Motorcycle tourism demand generators and dynamic interaction leisure. *International Journal of Culture, Tourism and Hospitality Research* 8(1), 92–102.

Tsiotsou, R.H. and Goldsmith, R.E. (eds) (2012) Target marketing and its application to tourism. In: *Strategic Marketing in Tourism Services*. Emerald Publishing Limited, Bingley, UK, pp. 3–15.

Uysal, M. and Hagan, L. (1993) Motivation of pleasure travel and tourism. In: Khan, M., Olsen, M. and Turgut, V. (eds) *VNR's Encyclopedia of Hospitality and Tourism*. Van Nostrand Reinhold, New York, pp. 798–810.

Walker, L. (2011) Tourism and leisure motorcycle riding. In: Prideaux, B. and Carson, D. (eds) *Drive Tourism: Trends and Emerging Markets*. Routledge, London, pp. 146–158.

Way, K.A., Roberts, L. and Turner, J. (2010) Travel and spending characteristics of the mature biker: An exploratory study. *Caesars Hospitality Research Summit* 12, 1–13. Available at: https://digitalscholarship.unlv.edu/hhrc/2010/june2010/12/ (accessed 14 October 2024).

Weddell, M. (2011) Travel preferences of recreational motorcyclists. *Journal of Tourism Insights* 5(1), 4.

Wheeller, B. (1991) Tourism's troubled times: Responsible tourism is not the answer. *Tourism Management* 12(2), 91–96.

Yoon, Y. and Uysal, M. (2005) An examination of the effects of motivation and satisfaction on destination loyalty: A structural model. *Tourism Management* 26(1), 45–56.

Zhang, X., Song, H. and Huang, G.Q. (2009) Tourism supply chain management: A new research agenda. *Tourism Management* 30(3), 345–358.

Part II: Resident Perspectives

This second part of the book consists of two chapters examining residents' perspectives of niche tourism. The first chapter of this part (Chapter 4), authored by Fani Efthymiadou and Chryso Panayidou, examines the perceptions of the residents of the island of Tilos (Greece) and their support for a newly implemented zero waste management initiative in the context of ecotourism. Emrullah Erul and Furkan Atasoy Karacabey authored the next chapter (Chapter 5), which investigates how the community attachment of the residents of Manavgat (Türkiye) and their perceptions of mass tourism influence their attitudes towards niche tourism development.

4 Assessing Resident Perceptions of Destination Ecotourism Image and Their Support of Environmental Practices

Fani Efthymiadou[1]* and Chryso Panayidou[2]

[1]Faculty of Tourism Administration, Boğaziçi University, Istanbul, Türkiye;
[2]Department of Tourism Management and Hospitality, Cyprus University of Technology, Pafos, Cyprus

Abstract

Niche tourism includes different types of tourism that focus on sustainability. One example of niche tourism is ecotourism, which is considered highly associated with the environment. In the ecotourism context, the development of direct environmental practices aims to conserve the environment. To that extent, the concept of zero waste management offers environmental benefits not only to the local community but also to the destination. This chapter focuses on resident perceptions over destination ecotourism image and examines their support in relation to the adoption of environmental practices by the local community. Drawing from interviews performed with residents of the island of Tilos (Greece), the findings indicate a high level of environmental awareness towards the adoption of a zero waste programme that is considered part and parcel of ecotourism development efforts. The support from the local authorities played an important role in enhancing resident awareness. Despite different stakeholders' conflict of interests, residents in Tilos support the zero waste programme. This chapter, therefore, offers several theoretical and practical implications which may contribute to the efficient development of ecotourism as a sustainable niche tourism form.

Keywords: Ecotourism, zero waste, resident perceptions, sustainability, governance

Introduction

Niche tourism refers to a specialized form of tourism that targets certain interests and preferences of tourists, but also promotes unique experiences and activities that are frequently in the form of alternatives to traditional travel options like mass leisure tourism. When approached with an emphasis on sustainability, niche tourism can benefit the environment and local communities (Howitt and Mason, 2018; Stronza *et al.*, 2019). Niche tourism developed within the framework of sustainability can promote conservation, cultural preservation and community involvement while aiming to reduce adverse effects on an area and its inhabitants (Mowatt, 2012). One such form of niche tourism is ecotourism, which represents one of the fastest-growing niche tourism markets. Ecotourism, centred in natural areas, is highly associated with environmental conservation, suggesting in essence that tourism should be

*Corresponding author: fani.efthymiadou@bogazici.edu.tr

DOI: 10.1079/9781800626669.0004

planned and managed under this guideline (Stronza *et al.*, 2019). In the context of ecotourism, the concept of zero waste management in tourism destinations becomes imperative as it aims not only to eradicate the problem of waste, but also to foster – through environmentally friendly practices – the concept of sustainable development. In this context, resident perceptions and support of ecotourism are important for its efficient development, a prerequisite being the implementation of environmental practices at the destination. Although resident perceptions have been previously studied in relation to ecotourism development, most studies focused on developing destinations in the African continent including Nigeria (Orimaye, 2018) and Botswana (Moswete and Thapa, 2015). The need to examine resident perceptions of Western destinations' ecotourism image and, by extent, their support in relation to the adoption of environmental practices such as zero waste is imperative.

Therefore, the aim of this chapter is to investigate resident perceptions of destination ecotourism image and their support in relation to environmental practices adopted at the destination. In so doing, we draw from resident perceptions of the island of Tilos in Greece. The study's focus on ecotourism at an island destination provides an interesting context that may yield significant implications and advance the debate on sustainability and niche tourism. The study has adopted a qualitative methodology using social exchange theory (SET), which is considered appropriate for research on resident perceptions and tourism support. It is considered a useful instrument in examining local attitudes towards tourism because of its versatility but also community engagement (Stergiou and Farmaki, 2020). The study involved informal interviews with residents of the island as well as in-depth interviews with local authorities. The data collected was analysed thematically to identify various themes associated with sustainable practices and destination images based on resident perceptions and expressions of support.

The findings revealed the residents' high level of interest in preserving their land and spreading awareness of their practices. Based on the mobilization and active participation of the local community, all waste generated on the island but treated in an environmentally friendly way could drastically reduce the impact from waste on the natural environment of Tilos. By doing so, the destination's ecotourism image may be enhanced by introducing sustainable practices. The zero waste programme might create a positive destination ecotourism image and attract tourists with environmental concerns who approve of these practices. The findings thus offer theoretical and practical contributions. Specifically, they offer directions for future sustainable practices that link to embedded residents' involvement in sustainable tourism development. Directions of how to achieve local support from authorities are also portrayed in this chapter, which concludes that governments, policy makers and tourism destination planners should further analyse sustainable practices in depth if they want to adopt similar strategies.

Theoretical Background

Ecotourism has the potential to significantly contribute to sustainability by protecting the environment, raising environmental consciousness and assisting local people. It requires planning, oversight and community involvement that are necessary for its success. Destinations need to make sure that tourism benefits the environment and local communities by upholding the principles of ecotourism and tackling its obstacles (Rinzin *et al.*, 2007).

Ecotourism and zero waste

In the travel industry, ecotourism has been recognized as an emerging paradigm. Stronza *et al.* (2019) define ecotourism as responsible tourism that conserves the environment, sustains the well-being of the local people and incorporates education. Establishing a positive relationship between tourism, environmental preservation and local community empowerment is one of the objectives of ecotourism. According to Fennell (2015), ecotourism goes beyond leisure travel and is ecologically consciously and actively involved in and supportive of local communities, placing a high emphasis on environmental preservation and protection. Furthermore,

Ceballos-Lascuráin (1996) argued that ecotourism emphasizes local community empowerment, environmental education and ethical travel activities. Lengieza *et al.* (2023) supported that the role of the local community shapes visitors' pro-environmental behaviour and concern for preserving the environment. According to Yoon and Lee (2023), ecotourism knowledge has a positive effect on residents' supportive attitudes towards ecotourism. This is further supported by Honey's (2018) research on ecotourism destinations in Latin America and Africa, which contends that the rapidly changing global tourism scene highlights the importance of ecotourism. Ecotourism is therefore a sustainable form of tourism that actively promotes the interests of the local community and enhances the environment. It serves as a road map for how environmentally conscious tourism may advance the goals of the community's environmentalists. By combining technology and community-based strategies, ecotourism may be utilized as a tool for sustainable development. For example, information and communication technology may promote tourism resources, increase the tourist arrivals or be the triggering event to develop new products (Marschall, 2012).

Ecotourism, as a sustainable niche tourism form, embraces the concept of zero waste. A viable path for sustainable tourism is presented by the incorporation of zero waste concepts into ecotourism activities and engaged community involvement. Zero waste is an all-encompassing strategy that reduces, reuses and recycles rubbish in an effort to lessen the negative effects of human activity on the environment. When used in reference to ecotourism, it refers to methods that aim to eradicate or manage waste produced by visitor activities in an ethical manner (Cobbinah, 2015). Zero waste practices can have an impact on the success of ecotourism. According to Rinzin *et al.* (2007), by adopting and implementing zero waste practices, ecotourism can have a positive effect on both the environment but also on community welfare. For example, an ecotourism initiative that contributes to sustainability takes place in the Galapagos Islands that offers guided tours with emphasis on biodiversity. Visitors are educated about the fragile ecotourism system by expert naturalists (Galápagos Conservancy, 2024).

Local community involvement in sustainable practices through ecotourism

The involvement of the local community in ecotourism can have a positive impact on the success of sustainable tourism initiatives. When the local community is more actively involved in decision making, favourable results for that community are more likely to occur (Phelan *et al.*, 2020). For instance, a balance between economic prospects and cultural preservation can be achieved and this can even ensure long-term sustainability for the destination. The degree to which local communities participate and engage with their environment is critical to the success of ecotourism. Current research, as demonstrated by studies like Moswete and Thapa (2015) and Smith and Jones (2022), emphasizes that community engagement entails more than just inclusion; it also involves benefit sharing, active participation in decision-making processes and a sense of ownership. Brown *et al.* (2021) argued that well-managed ecotourism can make a substantial contribution to local economies. This can be achieved through the formation of community-owned businesses, small-scale entrepreneurship and employment generation, which are recognized as important avenues.

Participation from local communities is also an important aspect of zero waste ecotourism. According to Jamal and Stronza (2009), local communities can play a crucial role as collaborators in the design, implementation and assessment of waste management methods and are not just passive hosts. Other researchers such as Hunt *et al.* (2014) also mention the potential of waste reduction strategies for ecotourism. Particularly, ecotourism work can deter single-use plastics that are not considered sustainable and this can be achieved through strict packaging regulations. Another ecotourism initiative by the local community is that of composting programmes (Lee and Wang, 2021), which can control organic waste but also reduce the possibility of waste being thrown into open areas. Zero waste strategies are important for ecotourism as they can promote material reduction, reuse and recycling as ways to preserve natural resources. This is particularly essential in ecotourism locations where the surrounding ecotourism serves as a major draw.

Garcia and Martinez (2022) mentioned that local community participation through local companies can also result in ecotourism development through the production of goods made by recycled materials that can further impact the community's financial stability. Ecotourism can lead the way to a more sustainable and responsible future by encouraging waste reduction practices, even if obstacles are evident like, for example, limited infrastructure and inequalities in waste management infrastructure between locations that are aiming for zero waste (Buckley, 2020). For this reason, Mostafanezhad and Hannam (2019) argued that community participation should involve open communication and teamwork, and should be addressed so that any obstacles are removed. Based on the above, our research aim is to investigate how locals perceive the ecotourism image of a destination and support environmental practices undertaken as part of zero waste management. In order to evaluate resident perceptions, we adopt SET which is widely used in examinations of resident perceptions and support (Farmaki *et al.*, 2019).

Social exchange theory and resident perceptions

As viewed through the prism of SET, the success and sustainability of ecotourism initiatives are claimed to be impacted by local community participation. SET is a theory developed by Blau (1964) and Emerson (1976). It states that in social interaction, individuals trade resources in an effort to maximize benefits and avoid any drawbacks. This theory provides a framework for understanding the dynamics in the perception of the benefits and costs by which local community decisions are made. In the context of tourism, the application of SET implies that residents will have positive perceptions and extend support for tourism development if they think the benefits of tourism outweigh the costs and vice versa. In light of the criticism surrounding mass tourism's negative impacts on destinations, SET emerges as a popular approach in studying resident perceptions and support of tourism. Nunkoo and Gursoy's (2017) study incorporates SET into sustainable tourism, emphasizing that locals' opinions of the economic, social and environmental exchanges linked to tourism development have a substantial impact on their support.

Hunt *et al.* (2014) highlighted the economic benefits that ecotourism offers to the local community. Garcia and Martinez (2022), on the other hand, discussed the potential cultural costs associated with ecotourism such as commercialization of culture and loss of national identity. However, Nguyen and Patel (2020) contended that when local residents actively engage in benefit sharing and decision-making processes, a more balanced power dynamic emerges, and this in turn fosters a healthy social exchange between residents and tourism. It becomes clear that the community-based ecotourism (CBET) model is consistent with the ideas of SET. In order to create a fair and reciprocal social exchange between other stakelolders, local communities actively participate in the planning, management and economic advantages of CBET, as demonstrated by Rodriguez and Garcia (2023).

Study Context: Tilos Island

Tilos is one of the smallest islands of the Dodecanese group of islands of Greece and is situated between the islands of Rhodes and Kos. According to the Hellenic Statistical Authority (2021), Tilos has only 746 inhabitants. Despite the small size of the island and the limited number of residents, Tilos has attracted national and international interest due to the environmental practices adopted by the island's authorities. Well-known for its unspoiled and beautiful nature, over the last few years it has been trying to promote sustainable development by the establishment of different policies such as energy sources that are 100% self-sufficient. Moreover, it is home to a single wind turbine, located in a remote site on Akri Punta Cape.

In 2021 Tilos became the first island in the world to abolish landfill and implement a waste management programme with the project name of 'Just Go Zero, Tilos'. The project runs under the accreditation of the region of Dodecanese and the supervision of the private company Poligreen. A circular economy centre has been built where waste is collected and sorted into three categories: organic waste, recyclables and non-recyclables. In addition to that, an application has been created to keep records for each resident individually in terms of the extent they contribute to the project and monitor their output of waste, which is transformed into new converted material. This project also involves tourists and especially evaluates their active participation during their stay (Just Go Zero, 2021).

Methodology

The study aimed to carry out:

1. An investigation into how the local residents perceive environmental practices.
2. An identification of awareness of local residents towards sustainable practices.
3. An exploration of the perceptions of the local residents between sustainable practices and the destination image through the impact of the zero waste programme.
4. An evaluation of how the residents might support or oppose the development of environmental practices.

Since the goal of the study was to obtain deep understanding of resident perceptions on Tilos' ecotourism image and their support of the zero waste programme implemented on the island, qualitative research was considered appropriate. The use of qualitative research is appropriate when seeking in-depth understanding of a situation (Korstjens and Moser, 2017). This approach explores a wide array of dimensions, understandings and experiences of research participants (Denzin *et al.*, 2017), and it seeks to understand and explore (Nassaji, 2020) phenomena under study.

Data collection

The interviews were performed with two groups of research informants regarded as representative of the residents of the destination. The first group included residents of the island and the second included participants who represented local authorities responsible also for tourism decision making and governance. In this way, a wider spectrum of information was obtained. Participants belonging to the first group were randomly selected using snowball sampling, where the researchers first identified residents willing to participate in the study who were then asked to suggest other potential participants. The second group of informants included local authority stakeholders chosen for their position and knowledge of the topic. Overall, eight participants were interviewed. Their profile can be seen in Table 4.1.

Prior to the interviews, the interviewees were briefed about the study and were assured that the data and information collected would be confidential. Interviews were conducted via Skype between February and March 2023. The duration of each interview was approximately 40min. Interviews

Table 4.1. Profile of the participants.

Profile of participants	Representation	Age (years)	Marital status	Gender
Local authorities				
Interviewee 1	Mayor	55	Married with children	f
Interviewee 2	Deputy mayor	49	Married with children	f
Local residents				
Interviewee 3	Local – employee	50	Divorced	f
Interviewee 4	Local – business owner	61	Single	m
Interviewee 5	Local – pensioner	67	Married with children	m
Interviewee 6	Local – housewife	65	Married	f
Interviewee 7	Local – fisherman	63	Divorced	m
Interviewee 8	Local – housewife	50	Married	f

were recorded with the permission of participants and were conducted in Greek and subsequently translated into English. First, the local participants were asked questions about their demographic profile such as their occupation, marital status and age. After this, the participants were asked to describe the zero waste programme, its relation to ecotourism but also how it can have an effect on sustainability. Moreover, they were asked how they perceived the role of the local community. Another theme category focused on to what extent the local authorities contributed to the application of the zero waste programme and what educational initiatives took place. Finally, the participants were asked to comment and generally review the programme of zero waste management. For the local authorities participants, the researchers slightly differentiated the questions in order to make them applicable to their individual circumstances. For example, questions about the role of local authorities were excluded and instead questions regarding the role of the local community were emphasized during their interviews. The information collected from the participants was analysed thematically (Rishi and Gaur, 2012) which means that various themes within the interview transcripts were identified with the aim of identifying patterns among them. The themes initially identified were subsequently further refined to produce sub-themes.

Findings

The main themes derived from the data analysis include: (i) environmental awareness – sustainable practices; (ii) community engagement and participation in the zero waste programme; (iii) the destination ecotourism image through the impact of the zero waste programme and the impacts of the programme; and (iv) evaluation of how the residents might support or oppose the development of environmental practices.

Environmental awareness and sustainable practices

The residents expressed the view that there is a need to implement the zero waste programme as a way to safeguard the environment. This would show that people are becoming more conscious of environmental problems and that there is a need to take preventative action to lessen any harm. The quotes below are evident of such perspectives:

> I support the progamme. I think is very good. People find this island very clean and beautiful. This is the image the tourists have of the island. Is a good system as they come and pick the rubbish. I originally come from Denmark so I am used to recycling and this is not new to me. In the summer time they go every day and even twice a day to hotels and restaurants to pick up the rubbish so is a really good programme.
>
> (Interviewee 6)

> I have positive thoughts; I am very happy as a local, as a citizen of this island. I had the chance to participate in a sustainable development pilot programme which turned my island into a shining global example. I am very proud to belong to such a community.
>
> (Interviewee 1)

The local municipality plays a key role in introducing and supporting the zero waste initiative. This demonstrates how crucial policy support and leadership are to the advancement of sustainable community-level projects. As Interviewee 4 said, 'Yes, they went from door to door and explained to every single resident what the programme is about and how to involve the local residents in this. And the result was good because all residents understood and got involved immediately'. Similarly, Interviewee 2 stated that 'The whole community has cooperated perfectly, we [as community] have adopted a responsible environmental behaviour probably because we are a small community and we have received a virgin, pure island, and we also have an awareness that we should preserve it and develop more regarding this matter'.

Local authorities' efforts to inform and include locals are credited with the programme's effectiveness. Effective programme participation is ensured for all inhabitants through door-to-door clarification in such a way that residents understand the programme well. 'There were people from the municipality who went door to door and explained to the residents what they should do, how they should manage and separate the waste. There was a lot of support from the authorities', said Interviewee 6. Interviewee 1 also posited that:

> The educational initiatives were the main reason to support the zero waste programme. We had several meetings at the main square, at schools, at military camps (we have a green military camp also). We also visited some residents' homes, especially older people, where they have accessibility issues. There was no household or business left that we did not visit and inform them about the way they should manage their waste.

Community engagement and participation

The community is crucial to the implementation and upkeep of the zero waste initiative. Residents have a shared sense of accountability and collaboration to accept and embrace the programme's principles and incorporate them into their everyday lives. The findings highlight the residents' active participation in the zero waste initiative, and underline their part in sorting and organizing rubbish. The fact that the island's population is comprised of a small community means that it is a manageable task to include and incorporate every inhabitant in the programme. In the words of Interviewee 8, 'I support the programme. I think it is very good. People find this island very clean and beautiful. This is the image that the tourists have of the island. It is a good system as they come and pick up the rubbish'. Interviewee 1 also said that 'this island has a philosophy and an orientation towards sustainable development, a focus on the protection of the environment; the island has adopted an ecological culture'. In agreement, Interviewee 2 argued, 'The locals were very supportive. I can say that locals are the ones who always are pushing for the implementation of such projects. Every programme we discussed among our local people for the environment, they always supported us'.

Potential for sustainability and a positive destination ecotourism image

With an emphasis on the long-term advantages for both current and future generations, the zero waste programme is considered as a means of averting harm before it arises. In the words of Interviewee 4, 'It keeps the environment clean and the tourists that come here see that the locals are involved. For sure it keeps the place clean and since they pick the rubbish up we don't see rubbish around on the island'. The interviewee went on to state, 'We know that the majority of European tourists are environmentally conscious ... for example, they are positively associated with a pioneering waste management system; as a result this acts as a motivation to visit our island and live the experience'.

These testimonies emphasize how crucial it is to implement sustainable practices in order to reduce the negative impact on the environment. 'At first it was an impossible mission; later I considered it very hard to accomplish, I had doubts how we can apply this. Also, the cost was huge until the private company funded the project', said Interviewee 2. Some residents, however, are sceptical or even suspicious of programmes like the zero waste initiative because they believe they are just surface-level solutions that do not deal with what causes the problem. According to one resident, these initiatives are motivated by convenience, but also more by political considerations than by a sincere desire to protect the environment and consequently ecotourism practices such as the zero waste programme. In particular, Interviewee 7 underlines:

> The zero waste programme is completely not necessary and does not help the island. It was not the actual mayor that initiated it but the central government. The local municipality just took the chance to show that they are doing something.

Perception of positive or negative impacts on the effectiveness of the zero waste programme

From the participants' voices it was indicated that the zero waste programme may be effortlessly incorporated into daily activities, irrespective of their personal or work-related conditions. This illustrates how flexible and adaptive the programme is in meeting the needs of different lifestyles. The initiative's value is recognized by the maintaining of an environment free from rubbish, and this can have an impact on the island's overall image in attracting tourism. This is turn could have a positive impact on ecotourism practices. This is another indication of the programme's apparent effect in keeping the island clean, and the way it is valued by the local residents. Residents recognize the significance of their current activities and actions in establishing a sustainable future for the island and perform ecotourism activities. This was emphasized through the responses of interviewees. For instance, 'It is clearly a positive impact. We have always supported ecology; for example, we have banned hunting, and this reversed the negative impact into a positive one', said Interviewee 1.

None the less, there was also a participant who disapproved of the local authorities' handling of urgent problems, including the development of the island's infrastructure (such as the sewage system) and economic assistance. Interviewee 7 stated, 'The island does not have a sewage system and the dirty waters go on the mountain and end up in the groundwater'.

Discussion

Sustainable practices in niche tourism include those that help local communities protect the environment but also conserve cultural heritage. The involvement of the local community is essential to the viability of any tourism development project (Wang *et al.*, 2021) but also to any form of environmental practices. Through the inclusion of local people in the planning, decision making and sharing of benefits, niche tourism products can promote cultural exchange, support sustainable development and practices, and improve the image of a destination in the overall travel experience (Nugroho and Numata, 2022). Indeed, this argument is in line with Kia (2021), who supports that a knowledge gap between local authorities and local people leads to unsuccessful participation by locals.

Overall, the present findings support the premise that promoting environmental protection depends largely on the engagement or active participation of the local community in ecotourism, and particularly the zero waste initiative. Nguyen and Patel's (2020) research emphasizes the significance of involving local populations in any efforts on waste reduction; it calls for customized solutions that are in line with the principles of a society and the cultural morals of a place. This, in turn, can also improve the general sustainability of tourism destinations and promote ecotourism initiatives. Residents show a collective commitment to environmental conservation and the zero waste programme, even in the face of personal conflicts with local officials. This finding is contrary to previous research, which identified conflictive interests of tourism stakeholders characterized as a barrier to achieving consensus (Alvarez, 2024).

The findings presented in this chapter reveal a strong feeling of shared identity among the community and a commitment to environmental care by extending the development of an ecotourism framework. Information sharing and education are emphasized as essential to the programme's success. It is acknowledged that for waste reduction measures to be widely adopted, education, information and advice are necessary. Moreover, according to Brown *et al.* (2021), if these are witnessed by visitors or guests at a destination, the visitors and guests can become supporters for zero waste practices in their own destinations. For example, information campaigns, guided tourism or even interactive workshops can help construct knowledge, change perceptions and encourage responsible waste reduction (Rodriguez and Garcia, 2023).

Last, it must be mentioned that there is a difference between the notion and image of a 'green island' and the harsh reality of resource mismanagement and environmental destruction. This contradiction highlights significant difficulties in accomplishing sustainability objectives such as the development of ecotourism in the face of real-world limitations and governance weaknesses. Evidently, there is a constant need to address resource mismanagement and environmental destruction (Blomley *et al.*, 2019) if sustainability is to be promoted in an ecotourism context.

Conclusion and Implications

This chapter has collated various theoretical contributions to the academic examination of ecotourism and reported a study that has made an evaluation of the stance of the local community in sustainable practices through the zero waste initiative, based on the perceptions of residents. The prevailing theory used to evaluate local perceptions towards the zero waste initiative has been SET. This theory affirms that residents evaluate the programme on the basis of the benefits and costs gained (Látková and Vogt, 2012). The study findings maintain the appropriateness of this theory as a theoretical framework that could ultimately shape a local community's views and perceptions of sustainable practices.

The projected input of the presented study was to fill major gaps in the literature by offering an approach on sustainable practices as seen from the perceptions of residents of an island destination. Hence, the acknowledged limited research on such initiatives provided an opportunity to consider the significance of ecotourism from an island destination perspective. The study findings are constructive for academia, through expanding and enlightening the breadth of information and knowledge of resident perceptions towards sustainable and environmental practices. The current study also offers practical implications to different stakeholders. For instance, important considerations are suggested to governments, policy makers and tourism destination planners, while also providing directed practical knowledge that acts as a guideline to the achievement of successful sustainable practices.

As this research is based on the island destination of Tilos in Greece, further research should be performed in other island destinations in order to reveal if there are any patterns among local communities that live in an island destination context. The projected programme on Tilos island is taking place through the commitment of a small group of people. This creates a limitation for the survey, so more research should be done using bigger sample data and a quantitative approach in order to examine the concept of applying ecotourism based on a zero waste programme on a large scale. The zero waste programme is a tool to promote ecotourism and sustainability; future research should thus consider different types of actions that promote sustainability. Certainly, in the tourism industry, more research on good case examples of successful governance in such contexts is needed.

References

Alvarez, M.D. (2024) Governance for the implementation of the sustainable development goals in tourism: A 2050 horizon paper. online version ahead of print, *Tourism Review*. DOI: 10.1108/TR-12-2023-0872.

Blau, P.M. (1964) *Exchange and Power in Social Life*. Wiley, New York.

Blomley, T., Iacona, G.D. and Boles, W.E. (2019) A critical analysis of the Australian Government's approach to offsets. *Environmental Management* 64(1), 1–11

Brown, A., Garcia, M. and Rodriguez, E. (2021) Reducing tourism's carbon footprint: An analysis of ecotourism packaging policies. *Journal of Sustainable Tourism* 45(2), 123–140.

Buckley, R. (2020) Tourism and biodiversity: More significant than climate change. *Journal of Sustainable Tourism* 38(6), 691–707.

Ceballos-Lascuráin, H. (1996) *Tourism, Ecotourism, and Protected Areas: The State of Nature-based Tourism Around the World and Guidelines for its Development*. IUCN, Gland, Switzerland.

Cobbinah, P.B. (2015) Contextualising the meaning of ecotourism. *Tourism Management Perspectives* 16, 179–189.

Denzin, N.K., Lincoln, Y.S., MacLure, M., Otterstad, A.M., Torrance, H. *et al.* (2017) Critical qualitative methodologies: Reconceptualizations and emergent construction. *International Review of Qualitative Research* 10(4), 482–498. DOI: 10.1525/irqr.2017.10.4.482.

Emerson, R.M. (1976) Social exchange theory. *Annual Review of Sociology* 2(1), 335–362.

Farmaki, A., Christou, P., Saveriades, A. and Spanou-Tripinioti, E. (2019) Perceptions of Pafos as European Capital of culture: Tourism stakeholder and resident perspectives. *International Journal of Tourism Research* 21(2), 234–244.

Fennell, D.A. (2015) *Ecotourism*. Routledge, London.

Galápagos Conservancy (2024) Protect Galápagos. Available at: https://www.galapagos.org/ (accessed 16 June 2024).

Garcia, N. and Martinez, F. (2022) Circular economy in ecotourism: From waste reduction to community empowerment. *Sustainability* 14(2), 456.

Hellenic Statistical Authority (2021) Census results of population and housing. Available at: https://elstat-outsourcers.statistics.gr/census_results_2022_en.pdf (accessed 14 October 2024).

Honey, M. (2018) *Ecotourism and Sustainable Development: Who Owns Paradise?*. Island Press, Washington, DC.

Howitt, J. and Mason, C.W. (2018) Ecotourism and sustainable rural development in Pérez Zeledón, Costa Rica. *Journal of Rural and Community Development* 13(1), 67–84.

Hunt, C.A., Durham, W.H., Driscoll, L. and Honey, M. (2014) Can ecotourism deliver real economic, social, and environmental benefits? A study of the Osa Peninsula, Costa Rica. *Journal of Sustainable Tourism* 23(3), 339–357. DOI: 10.1080/09669582.2014.965176.

Jamal, T. and Stronza, A. (2009) Collaboration theory and tourism practice in protected areas: Stakeholders, structuring and sustainability. *Journal of Sustainable Tourism* 17(2), 169–189.

Just Go Zero (2021) Η Τήλος γιόρτασε την κυκλική οικονομία (Tilos celebrated circular economy). Available at: https://www.justgozero.com/h-tilos-giortase-thn-kuklikh-oikonomia/ (accessed 14 October 2024).

Kia, Z. (2021) Ecotourism in Indonesia: Local community involvement and the affecting factors. *Journal of Governance and Public Policy* 8(2), 93–105.

Korstjens, I. and Moser, A. (2017) Series: Practical guidance to qualitative research. Part 2: Context, research questions and designs. *The European Journal of General Practice* 23(1), 274–279.

Látková, P. and Vogt, C.A. (2012) Residents' attitudes toward existing and future tourism development in rural communities. *Journal of Travel Research* 51(1), 50–67. DOI: 10.1177/004728751039419.

Lee, J. and Wang, L. (2021) Community-based waste reduction initiatives in ecotourism destinations. *Tourism Planning & Development* 25(1), 45–60.

Lengieza, M.L., Hunt, C.A. and Swim, J.K. (2023) Ecotourism, eudaimonia, and sustainability insights. *Journal of Ecotourism* 22(1), 43–58.

Marschall, S. (2012) Sustainable heritage tourism: The Inanda heritage route and the 2010 FIFA World Cup. *Journal of Sustainable Tourism* 20(5), 721–736.

Mostafanezhad, M. and Hannam, K. (2019) Cultural commodification, tourism and the creation of the 'other' in Iran. *Tourism Management* 74, 122–133.

Moswete, N. and Thapa, B. (2015) Factors that influence support for community-based ecotourism in the rural communities adjacent to the Kgalagadi Transfrontier Park, Botswana. *Journal of Ecotourism* 14(2–3), 243–263.

Mowatt, R.A. (2012) Cultural tourism as serious leisure. *Annals of Leisure Research* 15(1), 1–18.

Nassaji, H. (2020) Good qualitative research. *Language Teaching Research* 24(4), 427–431. DOI: 10.1177/1362168820941288.

Nguyen, T. and Patel, S. (2020) Empowerment through participation: A longitudinal study of community engagement in ecotourism initiatives. *Journal of Community Tourism Research* 15(1), 45–60.

Nugroho, P. and Numata, S. (2022) Resident support of community-based tourism development: Evidence from Gunung Ciremai National Park, Indonesia. *Journal of Sustainable Tourism* 30(11), 2510–2525.

Nunkoo, R. and Gursoy, D. (2017) Residents' support for tourism: An identity perspective. *Annals of Tourism Research* 64, 157–170.

Orimaye, J.O. (2018) Perception of ecotourism impact in Ekiti state: A case study of Ikogosi Warm Spring. *Agriculture Extension Journal*. Available at: https://ssrn.com/abstract=3663020 (accessed 14 October 2024).

Phelan, A., Ruhanen, L. and Mair, J. (2020) Ecosystem services approach for community-based ecotourism: Towards an equitable and sustainable blue economy. *Journal of Sustainable Tourism* 28(10), 1665–1685.

Rinzin, C., Vermeulen, W.J.V. and Glasbergen, P. (2007) Ecotourism as a mechanism for sustainable development: The case of Bhutan. *Environmental Sciences* 4(2), 109–125.

Rishi, M. and Gaur, S.S. (2012) Emerging sales and marketing challenges in the global hospitality industry: A thematic analysis of customer reviews from the world's top two tourist destinations. *Worldwide Hospitality and Tourism Themes* 4(2), 131–149.

Rodriguez, E. and Garcia, M. (2023) Community based ecotourism: A model for balanced social exchange. *Journal of Sustainable Development* 30(5), 512–527.

Smith, A. and Jones, B. (2022) Re-imagining community participation in ecotourism: A framework for inclusive decision-making. *Journal of Sustainable Tourism* 38(1), 78–95.
Stergiou, D.P. and Farmaki, A. (2020) Resident perceptions of the impacts of P2P accommodation: Implications for neighbourhoods. *International Journal of Hospitality Management* 91, 102411.
Stronza, A.L., Hunt, C.A. and Fitzgerald, L.A. (2019) Ecotourism for conservation? *Annual Review of Environment and Resources* 44(1), 229–253.
Wang, M., Jiang, J., Xu, S. and Guo, Y. (2021) Community participation and residents' support for tourism development in ancient villages: The mediating role of perceptions of conflicts in the tourism community. *Sustainability* 13(5), 2455.
Yoon, S.-E. and Lee, K.-J. (2023) The effect of ecotourism knowledge on residents' attitudes in Otavalo, Ecuador: The knowledge theory of attitude–behavior consistency. *Journal of Hospitality and Tourism Insights* 6(1), 174–190.

5 Tourism's Hidden Gem: Niche Tourism Unveiled Through Resident Insights

Emrullah Erul[1]* and Furkan Atasoy Karacabey[2]

[1]Faculty of Tourism, İzmir Katip Çelebi University, İzmir, Türkiye; [2]Faculty of Tourism, Sivas Cumhuriyet University, Sivas, Türkiye

Abstract

Niche tourism has emerged as a pivotal strategy in mitigating tourism's negative environmental, economic and sociocultural impacts. Previous studies highlight the importance of niche tourism as a counterbalance to mass tourism, fostering sustainable tourism development. This chapter, utilizing the stimulus–organism–response (SOR) theory, investigates how residents' attachment to their community and perceptions of mass tourism influence their attitudes towards niche tourism development. Additionally, it examines how these attitudes predict their support for niche tourism and expectations of its potential benefits. Data was collected through quantitative research conducted in Manavgat, Türkiye, a region characterized by intense mass tourism development. Results indicated that strong community ties and negative perceptions of mass tourism significantly enhanced favourable attitudes towards niche tourism. Furthermore, residents' attitudes towards niche tourism development significantly predicted their support for niche tourism and their expectations of its benefits. The chapter highlights the importance of involving residents in sustainable tourism practices, which can lessen the negative effects of mass tourism and support sustainable niche tourism development. These findings provide valuable guidance for policy makers and practitioners in crafting sustainable tourism strategies that meet the needs and expectations of local communities.

Keywords: Niche tourism, residents, community attachment, mass tourism, attitudes, SOR theory

Introduction

Niche tourism has gained popularity in recent years as a developmental option that addresses issues associated with mass tourism, including overtourism (Bunghez, 2021). The variety of niche tourism products, such as ecotourism, rural tourism, dark tourism and space tourism, reflects contemporary global tourists' wide range of interests. Niche tourism promotes responsible tourism practices, minimizing negative tourism impacts and preserving the environment and culture in destinations (Novelli, 2005). However, it is important to recognize that not all niche tourism products contribute equally to sustainability. For instance, while ecotourism and rural tourism often emphasize environmental conservation and cultural preservation, other niche tourism forms like golf tourism may not be as sustainable due to their intensive use of natural resources. Consequently, niche tourism generally plays a role in fostering sustainability in tourism. Hence, it is crucial for tourism planning to meticulously consider the development of niche tourism, ensuring that it provides

*Corresponding author: emrullah.erul@ikc.edu.tr

DOI: 10.1079/9781800626669.0005

maximum benefits to a destination, its community and sustainable tourism development in general (Butler, 2019).

Residents are one of the critical stakeholders of tourism as they are directly affected by the impacts of tourism development, thus playing a crucial role in the success of sustainable tourism development (Erul *et al.*, 2020). Their perceptions of and support for sustainable tourism initiatives are essential, as they influence the implementation and acceptance of sustainable practices within the community (Erul and Woosnam, 2022). In examinations of resident perceptions, several scholars (e.g. Gursoy *et al.*, 2010; Rasoolimanesh *et al.*, 2015; Garau-Vadell *et al.*, 2018; Hateftabar and Chapuis, 2020) focused on economic aspects (perceived positive economic tourism impacts, individual economic benefits, etc.) while some (e.g. Woosnam, 2012; Li and Wan, 2017; Erul *et al.*, 2020, 2023; Erul and Woosnam, 2022) emphasized interpersonal emotional elements (commitment, emotions, feelings, solidarity, etc.) to measure residents' attitudes towards tourism development. However, to the best of our knowledge, no studies have examined the relationship between the negative impacts of mass tourism, residents' degree of community attachment and future niche tourism development to ultimately explain residents' support for niche tourism and expectations of potential niche tourism benefits.

In other words, using the stimulus–organism–response (SOR) theory, this chapter reports a study investigating how residents' perceptions of mass tourism and their attachment to their community affect their attitudes towards niche tourism development, and how these attitudes influence their support for niche tourism and expectations of its potential benefits. The study utilized quantitative research methods to collect and analyse data from Manavgat, Türkiye, a region known for its extensive mass tourism development. This case study highlights the importance of fostering strong community ties and addressing residents' negative perceptions of mass tourism to gain their support for niche tourism. The results contribute to the understanding of sustainable tourism practices by emphasizing the role of residents in promoting niche tourism, which can mitigate the adverse impacts of mass tourism and enhance the sustainability of tourism development. The results offer valuable insights for policy makers and practitioners aiming to develop sustainable tourism strategies that align with the needs and expectations of local communities.

Literature Review

Mass tourism is recognized as one of the most prevalent forms of tourism, widely chosen by numerous individuals due to its cost-effectiveness, the convenience and assurance of having travel arrangements made by a travel agency, and the opportunity for extroverted individuals to meet a wide array of people (Vainikka, 2015). However, it is important to remember that travel agencies can also provide opportunities for niche tourism products and alternative tourism. Specialized travel agencies offer tailored experiences such as ecotourism, adventure tourism, cultural tourism and more, meeting the diverse preferences and needs of travellers seeking unique and personalized travel experiences.

Despite its significance in tourism, its extensive popularity can exert substantial impacts, especially when considering small or emerging destinations (Vanhove, 1997). For example, mass tourism can result in money leakage from the host community, decrease residents' quality of life and diminish the interaction between hosts and guests (Dwyer and Thomas, 2012). These impacts can adversely affect the destination's sociocultural, environmental and economic carrying capacities (Brandolini and Mosetti, 2005). Consequently, it is deemed crucial to conduct studies to minimize these effects and create a situation conducive to the well-being of residents.

Previous studies reveal that residents have a prevalent negative perspective towards mass tourism development (Woosnam and Erul, 2017) because mass tourists often remain within their hotel premises, barely interact with the local population and contribute minimally to the local economy (Anderson, 2012). This scenario poses detrimental effects at the destination level and can lead to economic losses for the country (Kondo, 2008). Identifying and addressing these challenges through research is considered vital.

Accordingly, the following hypothesis has been formulated:

> *H1: Increasing awareness of the negative impacts of mass tourism on local businesses correlates positively with residents' favourable attitudes toward future niche tourism development.*

Residents' attitudes, intentions and behaviour have been the subject of extensive research by scholars over many years, investigated through various theories including place attachment, the theory of planned behaviour, social exchange theory and emotional solidarity among others (e.g. Allen *et al.*, 1993; Almeida García *et al.*, 2015; Olya *et al.*, 2018; Erul *et al.*, 2020, 2023; Han *et al.*, 2023). From all these theories and models, community attachment has attracted several scholars' attention and was used in numerous studies (e.g. Choi and Murray, 2010; Lee, 2013; Eslami *et al.*, 2019; Kim *et al.*, 2021) to explain residents' attitudes, perceptions towards tourism impacts and support for tourism development. Community attachment refers to the social, emotional and psychological connections individuals establish with other individuals within a community (Eslami *et al.*, 2019). In other words, it is a sense of belonging and connection residents feel towards their community, which significantly influences their perceptions of tourism impacts – both positive and negative – and their subsequent support for tourism initiatives.

Previous studies have examined the relationship between community attachment and resident attitudes towards tourism development in various contexts. Choi and Murray (2010) conducted their research in a rural community in Texas, focusing on rural tourism. They found a significant positive relationship between community attachment and perceived positive impacts, indicating that residents who feel a strong connection to their community are more likely to view additional tourism development positively. Kim *et al.* (2021) explored this relationship within urban settings in South Korea, addressing general urban tourism, while Li and Wan (2017) investigated community attachment in the context of festival development in Macao. Their results echoed those of Choi and Murray (2010), demonstrating that community attachment positively influences residents' perceptions of tourism impacts and attitudes towards tourism development.

Furthermore, Eslami *et al.* (2019) claimed that as the degree of community attachment increases, the benefits of tourism become more evident, particularly in the context of cultural tourism in Langkawi Island, Malaysia. Similarly, Lee (2013) found the perceived benefits mediating between community attachment and support for tourism development in the context of sustainable tourism in south-west Taiwan. Olya *et al.* (2018) found that community attachment significantly predicted residents' support for tourism development with a specific focus on heritage tourism in Iran. Similarly, Adongo *et al.* (2017) found that residents with a stronger degree of community attachment were more supportive of heritage tourism development in Hoi An, Vietnam. These studies collectively illustrate that community attachment plays a crucial role in shaping residents' perceptions and support for tourism development across different contexts and types of tourism.

Moreover, previous studies showed that residents' support for tourism was explained by their attitudes towards tourism development. For instance, Erul *et al.* (2020) found that residents' emotional solidarity with tourists significantly predicted their attitudes towards tourism development and, ultimately, this attitude was a significant predictor of their support for tourism in the context of sustainable tourism in İzmir, Türkiye. Based on this, the following hypotheses were formulated:

> *H2: The strength of residents' community attachment significantly predicts their attitudes towards future niche tourism development.*

> *H3: Residents' attitudes towards future niche tourism development significantly predict their support for niche tourism.*

In tourism, the term *niche* evolves into a form that caters to individuals' needs, offering a more personalized and beneficial experience distinct from mass tourism (Novelli, 2005). Niche tourism also provides several benefits. First, it fosters interactions between tourists and local communities that preserve local culture and enable meaningful communication (Ursache, 2015). Additionally, by addressing the specific needs of individuals, niche tourism enhances the tourist experience and increases satisfaction (Zhou *et al.*, 2021). In summary, niche tourism is a vital component of sustainable tourism, offering significant benefits like preserving resources, preventing the degradation of fauna, enriching local economies and enhancing tourist experiences (Butler, 2019). Thus, this research formulates the following hypothesis:

> *H4: Residents' attitudes towards future niche tourism development significantly predict their expectation of potential niche tourism benefits.*

Using Manavgat, Antalya, Türkiye as the study site, the study reported in this chapter adopts the SOR model to test the proposed hypotheses. The model provides a conceptual framework by describing how individuals process external stimuli internally and lead to observable responses (Jacoby, 2002). This model, widely recognized for its application in consumer behaviour analysis, dissects the factors that shape attitudes and subsequent actions in the marketing and psychological realms (Kim *et al.*, 2020). Briefly, SOR looks at the connection people have with their town (stimulus), considers what they think and feel about it (organism), and then sees how that influences their actions, like backing different tourism projects (response). By investigating these steps, this chapter explores how the negative impacts of mass tourism and the sense of community attachment determine future niche tourism development and, ultimately, explains how this construct (future niche tourism development) will influence their support for niche tourism and expectations of potential niche tourism benefits (illustrated in Fig. 5.1).

Methods

Antalya is located in the southern part of Türkiye and on the coast of the Mediterranean Sea. With its wealthy cultural, historical and natural beauty, it attracts the attention of millions of tourists every year. Furthermore, Antalya hosts many sports events and attracts approximately 10 million tourists annually (Woosnam and Erul, 2017). The sample population for the present study consists of the residents living in Manavgat, a vital tourist district of Antalya, Türkiye. Manavgat is particularly suitable for this study due to its significant role in regional tourism, the high potential of its all-inclusive resorts, its diverse tourist attractions and substantial resident engagement in tourism-related activities. These factors make it an ideal location to explore resident attitudes towards tourism development.

The study employed a quantitative research design to explore resident attitudes towards niche tourism development. The survey

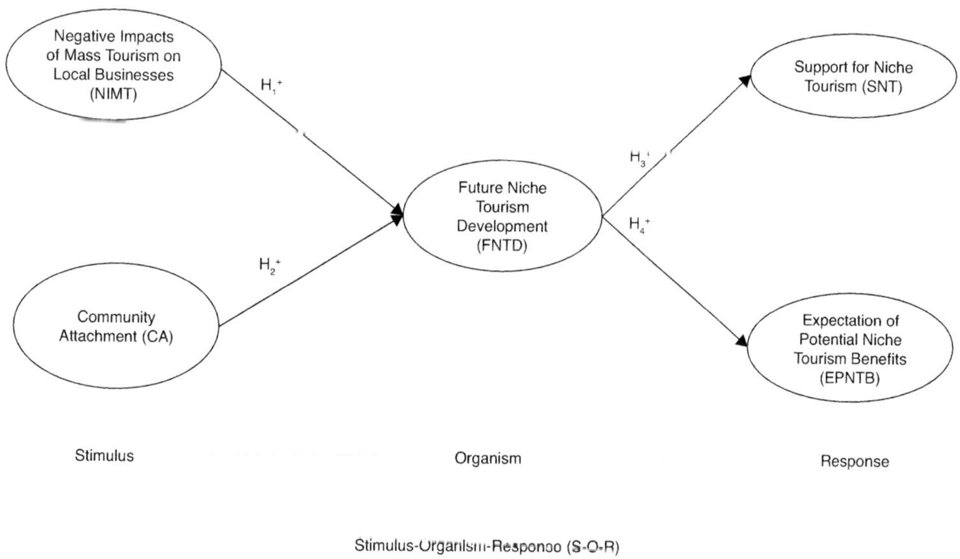

Fig. 5.1. The structural model.

questions were distributed through cluster sampling and were conducted voluntarily in 2021 at the end of the tourism season with individuals aged 18 years and over. Streets were randomly chosen using a map of Manavgat. On each of these selected streets, every fourth home or business was visited, and the head of the household or store employee was asked to participate. Researchers first explained the purpose of the study to potential participants, collecting data from those who agreed to participate. A total of 452 individuals were approached in the study, of whom 49 refused to complete the survey. Among the remaining participants, 42 either randomly marked the survey questions or left them incomplete, resulting in 361 participants being included in the study. The collected survey data were analysed using the statistical software packages SPSS 21 and AMOS 24 to test the conceptual model.

The survey questions were translated into Turkish by a professional translator and tourism researcher and then translated back into English from Turkish to prevent any loss of meaning. This research used a 5-point Likert scale ranging from 1 = 'strongly disagree' to 5 = 'strongly agree'. The study questions were adapted from previous studies in the literature on community attachment (Matarrita-Cascante *et al.*, 2006), negative impacts of mass tourism (Woosnam and Erul, 2017), future niche tourism development and expectations of potential niche tourism benefits (Erul *et al.*, 2023), and support for niche tourism (Erul, 2021). Confirmatory factor analysis (CFA) resulted in all factors being unidimensional or single.

Results

Sample characteristics

As seen in Table 5.1, male participants (n=187) and female participants (n=174) are almost in equal proportion (52% male and 48% female). More than half of the participants (54%) are involved in tourism-related professions. The majority of the sample is young (76%) and educated (57%) (with a total of 275 individuals aged between 18 and 39 years, while 177 individuals have at least a university degree). Nearly half of the participants (46%) are single (n=166).

Reliability and validity scores

As indicated in Table 5.2, the five factors' composite reliability (CR) scores range from 0.76 to 0.91. Both the Fornell–Larcker criterion (see Table 5.3) and the heterotrait–monotrait (HTMT) ratio of correlation (see Table 5.4) were included in the analysis of discriminant validity and the researchers found no issues or concerns. Briefly, all factors are considered reliable (CR ≥0.70) and valid (average variance extracted (AVE) scores > 0.50) (Hair *et al.*, 2019).

Fit indices

Model fit index values such as the Tucker–Lewis index (TLI), comparative fit index (CFI), incremental fit index (IFI) and root-mean-square error of approximation (RMSEA) were used to measure the goodness of fit of the measurement and structural models (Byrne, 2016). The fit index values for the measurement model and structural equation model (SEM), as shown in Table 5.5, are greater than the threshold value of 0.95, and the RMSEA value is also less than 0.50. These results indicate that the generated models have good fit index values (Byrne, 2016). All these values indicate that both the CFA and SEM fall within acceptable fit criteria (Hair *et al.*, 2019).

Hypothesis testing

In the study, the first two hypotheses (H1 and H2) test how the level of community attachment (CA) in tandem with residents' perception of the negative impacts of mass tourism on local businesses (NIMT) affect their attitudes towards the future niche tourism development (FNTD). NIMT is a significant determinant for FNTD (H1: β=0.24, t=5.96, P<0.001). Furthermore, there is a significant relationship between CA and FNTD (H2: β=0.30, t=6.92, P<0.001). Therefore, both hypotheses H1 and

Table 5.1. Demographic profile of the participants.

	n	%
Gender (*n*=361)		
Male	187	51.8
Female	174	48.2
Marital status (*n*=361)		
Single	166	46.0
Married	169	46.8
Divorced or separated	17	4.7
Widowed	9	2.5
Education (*n*=361)		
High school diploma or less	118	32.7
Associate degree	35	9.7
Bachelor's degree	177	49.0
Master's degree or higher	31	8.6
Employment (*n*=361)		
Not tourism-related	163	45.2
Tourism-related	197	54.6
Age (*n*=361)		
18–29 years	177	49.0
30–39 years	98	27.1
40–49 years	64	17.7
50–59 years	21	5.8
≥ 60 years	1	0.3

H2 are supported. Hypotheses H3 and H4 test the relationship between FNTD, the expectations of potential niche tourism benefits (EPNTB) and the support for niche tourism (SNT). As proposed in the study, FNTD positively and significantly influences both SNT (H3: $\beta=0.41$, $t=8.57$, $P<0.001$) and EPNTB (H4: $\beta=0.70$, $t=13.63$, $P<0.001$). Thus, hypotheses H3 and H4 are accepted, meaning all study hypotheses are accepted (see Table 5.6).

The analysis showed that all hypotheses put forward were statistically significant. Specifically, 49% of the variability in expectation of potential niche tourism benefits ($R^2=0.49$) and 17% of the variability in support for niche tourism was attributed to attitudes regarding future niche tourism development ($R^2=0.17$). The study results also indicate that both community attachment and negative impacts of mass tourism were significant determinants of future niche tourism development ($R^2=0.17$ means CA and NIMT explained 17% of the variability in FTND).

Discussion

The current study found that as the residents' perceived negative impacts of mass tourism increased, their attitudes towards future niche tourism development also improved. Similar to our results, previous scholars (Bartolomé *et al.*, 2009; Gursoy *et al.*, 2010; Dillette *et al.*, 2017; Woosnam *et al.*, 2019) established that residents' perceived impacts of tourism and their general perceptions of tourism shape their attitudes towards niche forms of tourism. For instance, Woosnam *et al.* (2019) found that residents' perceptions of tourism correspond to their attitudes

Table 5.2. Measurement model results.

Constructs and indicators	Factor loading	*t*-Statistic	CR	AVE
Negative impacts of mass tourism on local businesses (NIMT)			0.91	0.67
NIMT-1	0.93	N/A		
NIMT-2	0.91	37.88[a]		
NIMT-3	0.76	25.99[a]		
NIMT-4	0.65	20.03[a]		
NIMT-5	0.81	29.14[a]		
Community attachment (CA)			0.83	0.51
CA-1	0.78	N/A		
CA-2	0.80	20.96[a]		
CA-3	0.86	21.99[a]		
CA-4	0.51	12.74[a]		
CA-5	0.53	13.14[a]		
Future niche tourism development (FNTD)			0.87	0.52
FNTD-1	0.79	N/A		
FNTD-2	0.75	20.04[a]		
FNTD-3	0.76	20.28[a]		
FNTD-4	0.71	18.78[a]		
FNTD-5	0.67	17.45[a]		
FNTD-6	0.64	16.48[a]		
Support for niche tourism (SNT)			0.77	0.54
SNT-1	0.75	N/A		
SNT-2	0.87	15.20[a]		
SNT-3	0.53	12.43[a]		
Expectation of potential niche tourism benefits (EPNTB)			0.76	0.51
EPNTB-1	0.74	N/A		
EPNTB-2	0.78	16.21[a]		
EPNTB-3	0.62	13.86[a]		

N/A, not applicable.
[a]$P < 0.001$.

towards ethnic neighbourhood tourism as a form of niche tourism. Furthermore, Gursoy *et al.* (2010) concluded that residents who perceived the negative impacts of tourism strongly opposed mass tourism development.

Furthermore, this study directly examined the relationship between the residents' degree of attachment to their community and their attitudes towards niche tourism, therefore supporting previous studies (see Choi and Murray, 2010; Adongo *et al.*, 2017; Li and Wan, 2017; Kim *et al.*, 2021) which reported that community attachment is a significant predictor of residents' attitudes towards tourism development. For instance, Adongo *et al.* (2017) found that residents who are more attached to their community tend to have more positive or favourable attitudes towards tourism development.

Table 5.3. Discriminant validity analysis (Fornell–Larcker criterion).

Measures	EPNTB	CA	NIMT	FNTD	SNT
EPNTB	0.72	–	–	–	–
CA	0.38	0.71	–	–	–
NIMT	0.14	0.11	0.82	–	–
FNTD	0.69	0.31	0.28	0.72	–
SNT	0.30	0.15	0.24	0.40	0.73

Table 5.4. Discriminant validity analysis (HTMT ratio).

Measures	EPNTB	CA	NIMT	FNTD	SNT
CA	–	–	–	–	–
NIMT	0.10	–	–	–	–
FNTD	0.28	0.27	–	–	–
SNT	0.13	0.24	0.48	–	–
EPNTB	0.34	0.10	0.66	0.32	–

Table 5.5. Fit indices of the measurement and structural models.

	CMIN (χ^2)	Degrees of freedom (df)	CMIN/df	IFI	TLI	CFI	RMSEA
Measurement model	435.828	197	2.212	0.967	0.961	0.977	0.043
Structural model	468.385	202	2.319	0.963	0.957	0.963	0.045

Similarly, Choi and Murray (2010) concluded that residents with a strong attachment to their community tend to show more positive attitudes than those who are less attached. The present study contributes significantly to the literature by identifying that residents' attachment to the community is a meaningful precursor to their attitudinal support for niche tourism development.

Finally, the current study observed that residents' attitudes towards niche tourism development in turn significantly determined both their support for niche tourism development and their expectation of potential niche tourism benefits. These results aligned with the results of previous studies (Látková and Vogt, 2012; Liu *et al.*, 2018; Boukas, 2019; Erul *et al.*, 2020; Erul and Woosnam, 2022). For example, Erul and Woosnam (2022) concluded that residents' attitudes towards tourism development significantly influence their intentional support, ultimately explaining their actual support for tourism development in İzmir, Türkiye. In a similar vein, Látková and Vogt (2012) claimed that residents who had positive attitudes towards tourism development were more optimistic and supported future tourism development. Similarly, Boukas (2019) investigated the impact of rural tourism on residents' well-being, identifying rural tourism as a niche with the potential to enhance the island's competitiveness and improve residents' quality of life. However, in contrast to our results, Liu *et al.* (2018) found an opposite path between residents' attitudes towards niche tourism development and their expectation of potential niche tourism benefits. The authors concluded that expected benefits were significant predictors of their supportive attitudes towards nature-based tourism development.

Conclusions and Implications

This chapter reports a study that aimed to investigate how community attachment and the perceived negative impacts of mass tourism

Table 5.6. Hypothesized relationships.

Hypothesis	Unstandardized coefficient, *B*	Standardized coefficient, beta (β)	*t*-Statistic	Supported
H1: Negative impacts of mass tourism on local businesses (NIMT) → Future niche tourism development (FNTD)	0.18	0.24	5.96[a]	Yes
H2: Community attachment (CA) → Future niche tourism development (FNTD)	0.22	0.30	6.92[a]	Yes
H3: Future niche tourism development (FNTD) → Support for niche tourism (SNT)	0.38	0.41	8.57[a]	Yes
H4: Future niche tourism development (FNTD) → Expectation of potential niche tourism benefits (EPNTB)	0.76	0.70	13.62[a]	Yes

[a]$P < 0.001$; RMSEA: FNTD = 0.17, SNT = 0.17, EPNTB = 0.49.

influence residents' attitudes towards future niche tourism development and, ultimately, how these attitudes significantly and positively determine their support for niche tourism development and expectation of potential niche tourism benefits. By applying the SOR theory, we analysed the relationships between these variables using survey data collected from residents in Manavgat, Antalya, Türkiye. Our findings revealed that both community attachment and the perceived negative impacts of mass tourism were significant predictors of future niche tourism development. Specifically, strong community attachment and awareness of the adverse effects of mass tourism positively influenced residents' attitudes towards future niche tourism development. These attitudes, in turn, significantly determined their support for niche tourism development and their expectations of the benefits it could bring. These results underscore the importance of fostering community attachment and addressing the negative impacts of mass tourism to promote sustainable niche tourism development.

The current study offers several contributions to examine residents' support for niche tourism development. First, it supports the SOR model in explaining residents' support for niche tourism development. The results show that community attachment and perceived negative impacts of mass tourism significantly and positively influence residents' attitudes towards future niche tourism development (stimulus). In other words, those factors acted as precursors to attitudinal support for tourism development, consistent with the findings of Choi and Murray (2010), Adongo *et al.* (2017) and Woosnam *et al.* (2019). These attitudes (organism) explained a significant portion of the variance in both residents' support for niche tourism development and their expectation of potential niche tourism benefits (response). To the authors' knowledge, the present study is one of the first to utilize these frameworks together to understand what contributes to residents' support for niche tourism development in their community. Another significant contribution is that, although benefits have been used to predict residents' support (see Gursoy *et al.*, 2010; Liu *et al.*, 2018), research had not yet focused on whether residents' positive attitudes towards niche tourism development may explain their expectation of potential niche tourism benefits.

The study findings also have several practical implications for policy makers, government officials, managers and planners in Manavgat for sustainable niche tourism development. First, hosting niche tourism events, including seminars and fairs, may positively impact tourism development by improving residents' integration and fostering genuine interactions with tourists. For instance, in a destination that considers wine tourism as its niche, destination managers could organize a wine festival. These events will actively involve residents by providing opportunities for them to participate in planning and execution, offering local products and showcasing their culture. Moreover, these tourism activities and events will foster a sense of belonging within the local community, as residents take active roles in planning and executing the events and increase the benefits of niche tourism, promoting a more positive outlook towards niche tourism development.

Furthermore, policy makers should play a crucial role in actively engaging with residents when planning niche tourism development, considering their needs and concerns. To achieve this, weekly or monthly meetings should be organized and included in tourism planning. Importantly, policy makers should ensure that residents are given a strong voice and appropriate representation in projects supporting niche tourism development. The direct involvement of the local community in tourism-related meetings, activities, events and tourism developments plays a significant role in the sustainability of tourism (Zhang *et al.*, 2020). Moreover, government officials should create education sessions for residents to raise awareness about the positive and negative impacts of niche tourism and its contribution to society and the destination. These sessions are crucial in ensuring that the residents are well-informed about tourism development. Furthermore, conducting surveys among residents at regular intervals contributes to the destination's sustainability. This approach identifies when residents' support is present or absent, providing valuable insights for strategic planning and development. The success of these sessions and surveys relies on the active participation and understanding of the residents. Destination planners can organize these sessions and surveys, adjusting tourism strategies based on the insights gained. Industry practitioners can also collaborate with local communities by creating job opportunities, promoting local products and involving residents in tourism decision-making processes, thereby enhancing the overall tourism experience and fostering community support.

As with any scholarly endeavour, the study reported here is subject to certain limitations. The first limitation arises from the data being collected within a restricted time frame; additionally, the study's focus solely on quantitative data can be seen as a limitation. Future research is advised to pay attention to this aspect and adopt alternative or complementary research methods. Moreover, the study concentrates on a single destination. Future studies exploring various countries and different destinations within the context of niche tourism are needed as incorporating perspectives from diverse cultures would be intriguing. Lastly, focusing on specific niche tourism forms as well as targeting particular demographic groups like women, in line with Sustainable Development Goals (SDGs), may be promising.

References

Adongo, R., Choe, J.Y. and Han, H. (2017) Vietnam: Impacts, perceived benefits, community attachment and support for tourism development. *International Journal of Tourism Sciences* 17(2), 86–106.

Allen, L.R., Hafer, H.R., Long, P.T. and Perdue, R.R. (1993) Rural residents' attitudes toward recreation and tourism development. *Journal of Travel Research* 31(4), 27–33.

Almeida García, F., Balbuena Vázquez, A. and Cortés Macías, R. (2015) Resident's attitudes towards the impacts of tourism. *Tourism Management Perspectives* 13, 33–40.

Anderson, W. (2012) Analysis of 'all-inclusive' tourism mode in the Balearic Islands. *Tourismos* 7(1), 309–323.

Bartolomé, A., Ramos, V. and Rey-Maquieira, J. (2009) Residents' attitudes towards diversification sports tourism in the Balearics. *Tourism Recreation Research* 34(1), 55–65.

Boukas, N. (2019) Rural tourism and residents' well-being in Cyprus: Towards a conceptualised framework of the appreciation of rural tourism for islands' sustainable development and competitiveness. *International Journal of Tourism Anthropology* 7(1), 60–86.

Brandolini, S.M.D.A. and Mosetti, R. (2005) Social carrying capacity of mass tourist sites: Theoretical and practical issues about its measurement. In: *Second International Conference on 'Tourism and Sustainable Economic Development – Macro and Micro Economic Issues', Chia, Italy, 16–17 September 2005*. Available at: https://core.ac.uk/download/pdf/6264102.pdf (accessed 15 October 2024).

Bunghez, C.L. (2021) The emerging trend of niche tourism: Impact analysis. *Journal of Marketing Research and Case Studies* 134710.

Butler, R.W. (2019) Niche tourism (birdwatching) and its impacts on the well-being of a remote island and its residents. *International Journal of Tourism Anthropology* 7(1), 5–20.

Byrne, B.M. (2016) *Structural Equation Modeling with AMOS: Basic Concepts, Applications, and Programming*, 3rd edn. Routledge, New York.

Choi, H.C. and Murray, I. (2010) Resident attitudes toward sustainable community tourism. *Journal of Sustainable Tourism* 18(4), 575–594.

Dillette, A.K., Douglas, A.C., Martin, D.S. and O'Neill, M. (2017) Resident perceptions on cross-cultural understanding as an outcome of volunteer tourism programs: The Bahamian family island perspective. *Journal of Sustainable Tourism* 25(9), 1222–1239.

Dwyer, L. and Thomas, F. (2012) Tourism yield measures for Cambodia. *Current Issues in Tourism* 15(4), 303–328.

Erul, E. (2021) Residents' desire for potential tourism development: The case study of Antalya, Turkey. *Journal of Turkish Tourism Research* 5(3), 2200–2116.

Erul, E. and Woosnam, K.M. (2022) Explaining residents' behavioral support for tourism through two theoretical frameworks. *Journal of Travel Research* 61(2), 362–377.

Erul, E., Woosnam, K.M. and McIntosh, W.A. (2020) Considering emotional solidarity and the theory of planned behavior in explaining behavioral intentions to support tourism development. *Journal of Sustainable Tourism* 28(8), 1158–1173.

Erul, E., Woosnam, K.M., Ribeiro, M.A. and Salazar, J. (2023) Complementing theories to explain emotional solidarity. *Journal of Sustainable Tourism* 31(2), 229–244.

Eslami, S., Khalifah, Z., Mardani, A., Streimikiene, D. and Han, H. (2019) Community attachment, tourism impacts, quality of life and residents' support for sustainable tourism development. *Journal of Travel & Tourism Marketing* 36(9), 1061–1079.

Garau-Vadell, J.B., Gutierrez-Taño, D. and Diaz-Armas, R. (2018) Economic crisis and residents' perception of the impacts of tourism in mass tourism destinations. *Journal of Destination Marketing & Management* 7, 68–75.

Gursoy, D., Chi, C.G. and Dyer, P. (2010) Locals' attitudes toward mass and alternative tourism: The case of Sunshine Coast, Australia. *Journal of Travel Research* 49(3), 381–394.

Hair, J.F., Risher, J.J., Sarstedt, M. and Ringle, C.M. (2019) When to use and how to report the results of PLS-SEM. *European Business Review* 31(1), 2–24.

Han, S., Ramkissoon, H., You, E. and Kim, M.J. (2023) Support of residents for sustainable tourism development in nature-based destinations: Applying theories of social exchange and bottom-up spillover. *Journal of Outdoor Recreation and Tourism* 43, 100643.

Hateftabar, F. and Chapuis, J.M. (2020) How resident perception of economic crisis influences their perception of tourism. *Journal of Hospitality and Tourism Management* 43, 157–168.
Jacoby, J. (2002) Stimulus–organism–response reconsidered: An evolutionary step in modeling (consumer) behavior. *Journal of Consumer Psychology* 12(1), 51–57.
Kim, M.J., Lee, C.K. and Jung, T. (2020) Exploring consumer behavior in virtual reality tourism using an extended stimulus–organism–response model. *Journal of Travel Research* 59(1), 69–89.
Kim, S., Kang, Y., Park, J.H. and Kang, S.E. (2021) The impact of residents' participation on their support for tourism development at a community level destination. *Sustainability* 13(9), 4789.
Kondo, M. (2008) The economic impact of all inclusive resorts in Jamaica. *City Tech Writer* 3(2008). Available at: http://www.citytech.cuny.edu/english/docs/ctw_v3_2008.pdf (accessed 15 October 2024).
Látková, P. and Vogt, C.A. (2012) Residents' attitudes toward existing and future tourism development in rural communities. *Journal of Travel Research* 51(1), 50–67.
Lee, T.H. (2013) Influence analysis of community resident support for sustainable tourism development. *Tourism Management* 34, 37–46.
Li, X. and Wan, Y.K.P. (2017) Residents' support for festivals: Integration of emotional solidarity. *Journal of Sustainable Tourism* 25(4), 517–535.
Liu, S., Cheung, L.T., Lo, A.Y. and Fang, W. (2018) Livelihood benefits from post-earthquake nature-based tourism development: A survey of local residents in rural China. *Sustainability* 10(3), 699.
Matarrita-Cascante, D., Luloff, A.E., Krannich, R.S. and Field, D.R. (2006) Community participation in rapidly growing communities in Southern Utah. *Community Development* 37(4), 71–87.
Novelli, M. (2005) *Niche Tourism: Contemporary Issues, Trends, and Cases*. Elsevier, Amsterdam.
Olya, H.G.T., Alipour, H. and Gavilyan, Y. (2018) Different voices from community groups to support sustainable tourism development at Iranian world heritage sites: Evidence from Bisotun. *Journal of Sustainable Tourism* 26(10), 1728–1748.
Rasoolimanesh, S.M., Jaafar, M., Kock, N. and Ramayah, T. (2015) A revised framework of social exchange theory to investigate the factors influencing residents' perceptions. *Tourism Management Perspectives* 16, 335–345.
Ursache, M. (2015) Niche tourism markets – means of enhancing sustainable economic development in EU's eastern periphery. *CES Working Papers* 7(2A), 648–661.
Vainikka, V. (2015) Rethinking mass tourism: Professional discourses of contemporary mass tourism and destinations. *Nordia Geographical Publications* 44(2), 99.
Vanhove, N. (1997) Mass tourism: Benefits and costs. In: Wahab, S. and Pilgrim, J.J. (eds) *Tourism, Development and Growth: The Challenge of Sustainability*. Routledge, London, pp. 50–77.
Woosnam, K.M. (2012) Using emotional solidarity to explain residents' attitudes about tourism and tourism development. *Journal of Travel Research* 51(3), 315–327.
Woosnam, K.M. and Erul, E. (2017) Residents' perceived impacts of all-inclusive resorts in Antalya. *Tourism Planning & Development* 14(1), 65–86.
Woosnam, K.M., Maruyama, N.U., Ribeiro, M.A. and Joo, D. (2019) Explaining minority residents' attitudes of ethnic enclave tourism from general perceptions of tourism impacts. *Journal of Tourism and Cultural Change* 17(4), 467–484.
Zhang, Y., Chan, J.H., Ji, Z., Sun, L., Lane, B. *et al.* (2020) The influence of community factors on local entrepreneurs' support for tourism. *Current Issues in Tourism* 23(14), 1758–1772.
Zhou, J., Xiang, K., Cheng, Q. and Yang, C. (2021) Psychological and behavioural consistency value seeking of tourists in niche tourism: Nostalgia, authenticity perception, and satisfaction. *Psychology Research and Behavior Management* 14, 1111–1125.

Part III: Implementation Practices

This third part of the book includes four chapters examining niche tourism implementation practices, amid the challenges that destinations face in terms of sustainability. The first chapter of the part (Chapter 6) is authored by Athina Nella and Dimitrios Stergiou who analyse UK stakeholder practices in the context of birdwatching tourism, revealing varied sustainability efforts. The following chapter (Chapter 7), authored by Stella Kladou, Maria Alebaki and Kleio Arvanakou, draws from the island of Corfu (Greece) to explore sustainability practices and challenges for family wineries in the context of wine tourism. The next chapter (Chapter 8) by Sıla Karacaoğlu presents cases of creative tourism practices in Türkiye to examine the role of creative tourism in sustainable development of rural areas. The last chapter of this part (Chapter 9) is authored by Maria Hadjielia Drotarova and Vasilis Papavasiliou who explore the practices of small and medium tourism enterprises (SMTEs) in Cyprus in promoting sustainable rural tourism development.

6 Birdwatching Tourism: Perspectives, Impacts and Sustainability Considerations

Athina Nella* and Dimitrios P. Stergiou
School of Social Sciences, Hellenic Open University, Patras, Greece

Abstract
Birdwatching tourism, a niche within ecotourism, is recognized for its sustainability and economic benefits to local communities. This chapter explores the evolving market, provides information on birdwatcher profiles, and examines impacts on economies, societies and environments. In doing so, it highlights the dual role of birdwatching in promoting conservation and contributing to local economies while identifying challenges like wildlife disturbance and carbon emissions from travel. The chapter integrates a literature review and analysis of UK stakeholder practices, revealing varied sustainability efforts, ranging from eco-friendly travel options to community engagement initiatives. Addressing these challenges requires multifaceted strategies, including active participation from local communities, robust conservation laws, environmental education and leveraging technology. This approach aims to ensure that all stakeholders in birdwatching tourism both benefit from and contribute to its sustainable future.

Keywords: Avitourism, birdwatching tourism, sustainability, conservation, stakeholders

Introduction

Birdwatching, also known as birding, avitourism or ornithological tourism, has evolved from a recreational pursuit into a niche form of tourism within the ecotourism market (Gosálvez Rey *et al.*, 2022), demonstrating a promising outlook for the market's future (Future Market Insights, 2024). Within this context of growth, birdwatching is also widely regarded as a highly sustainable tourist activity (Ren *et al.*, 2022). Beyond its economic benefits on destinations and local communities, it fosters greater awareness of the importance of conserving natural resources, leading to increased investments in the conservation of natural heritage (Sánchez-Rivero *et al.*, 2020). This aspect of sustainability and its impacts position birdwatching tourism not only as a driver of community development but also as a contributor to sustainability efforts within the broader (eco)tourism industry.

Despite its significance, birdwatching tourism research is considered to be in its early stages, with a predominant focus on the economic aspects and demand-side dynamics (Steven *et al.*, 2015). Studies reveal significant gaps in understanding various aspects of birdwatching tourism, such as conservation at avitourism destinations and the birdwatching experience and behaviour (Steven *et al.*, 2015; Brock *et al.*, 2021). Considering these gaps and the expansion of avitourism, alongside growing concerns for sustainable tourism development, further research is needed to understand

*Corresponding author: a.nella@eap.gr

DOI: 10.1079/9781800626669.0006

patterns and sustainability practices within this niche market (Steven *et al.*, 2015; Xu *et al.*, 2023).

Against this background, we seek to contribute to the discourse on niche tourism and sustainability by examination of birdwatching tourism. More specifically, this chapter draws on a review of relevant literature and secondary data from desk research in terms of trends, impacts and sustainability initiatives of birdwatching tourism, so as to highlight key sustainability challenges and practices. This secondary data specifically includes a preliminary examination of website content from key birdwatching stakeholders in the UK. This examination aims to supplement the findings from the literature review and provide a contemporary perspective on market practices and sustainability issues.

Birdwatching Tourism, Market Prospects and Profiling of Birdwatchers

Concepts and dynamics of birdwatching tourism

Birdwatching tourism involves travel beyond one's usual environment to view birds in their natural habitat (Connell, 2009). This activity may go beyond the recreational enjoyment of observing birds with the use of equipment such as binoculars, cameras and telescopes, to engage birdwatchers in scientific classification, environmental issues and more detailed study of birds (Vas, 2017). As such, it is positioned as a highly sustainable form of nature-based tourism.

From a tourism perspective, enhancing the birdwatching experience is crucial, with birding quality and diversity along with the level of facilitation provided by specialized guides contributing significantly to satisfaction (Stemmer (formerly Jathe) *et al.*, 2022). The accessibility of birding destinations is supported from necessary tourism infrastructure, such as regular air transport and accommodation options, alongside field guides that help birdwatchers to compare their observations with existing literature (Werdler, 2017). Lee *et al.* (2010) have shown that increased service levels in birdwatching tours correlate with greater satisfaction among birdwatchers and a higher willingness to pay more for these experiences.

Specialized forms of birdwatching, such as twitching, ecological birdwatching and community-based avitourism, cater to specific needs within the birdwatching community. Twitching refers to the enthusiastic pursuit of spotting new or rare bird species (Connell, 2009), oftentimes requiring long-distance travel and significant investment of time and money (Brock *et al.*, 2021). Ecological birdwatching prioritizes minimal impact on bird habitats, aiming to observe without affecting their natural life (Liu *et al.*, 2021). Extending this ethos of minimal impact and conservation, community-based avitourism engages local communities who use their ornithological knowledge to guide and educate tourists, thereby generating supplementary income and supporting conservation efforts (Biggs *et al.*, 2011).

Alongside these specialized birdwatching practices, avitourism literacy also plays a crucial role in promoting sustainability and conservation. This encompasses the awareness, knowledge, values and behaviours that support the protection and improvement of bird habitats for current and future generations (Conradie and van Zyl, 2020). This literacy continuum spans from initial awareness and affinity towards birds to active engagement in conservation behaviours, demonstrating the levels of involvement required to support sustainable avitourism (Conradie and van Zyl, 2020).

Market overview and prospects

Birdwatching is becoming increasingly popular among European tourists, with the UK being the largest European source country (CBI, 2021). At the same time, emerging destinations from Asia, Latin America, the Caribbean and Africa are starting to prioritize avitourism. Africa is mentioned as the birding continent while South Africa has the largest number of regionally endemic bird species of any African country (Riley, 2014). Other popular birdwatching destinations include Costa Rica (over 900 bird species), the Galapagos Islands (over 2000 bird species), Kenya (over 1100 bird species), Peru (home to more than 15% of the world's bird species), the Falkland Islands (over 200 bird species), Andalusia and the Farne Islands (Grihault, 2023).

According to Future Market Insights (2024), the birdwatching tourism market is estimated to reach US$659.4 billion in 2024; with an average compound annual growth rate of 5.3% it is projected to reach US$1.1 trillion by 2034. The growth of the avitourism market can be attributed to increased numbers of bird enthusiasts and eco-friendly tourists, intense interest in nature-based activities, wildlife conservation efforts and the growing appeal of sustainable tourism practices (MarkWide Research, 2024). This economic expansion is not only fuelled by direct travel expenses but also by significant investments in specialized equipment. Understanding the full economic impact of birding, therefore, requires assessing both trip-related and equipment expenditures. Moreover, digital advancements, such as bird identification applications and virtual birdwatching experiences, have the potential to enhance the birdwatching tourism experience and create additional spending. For example, platforms like https://fatbirder.com and https://ebird.org (both accessed 15 October 2024) can assist birders in designing and enriching their birding experiences. The increasing popularity of birdwatching tourism highlights its significant potential for growth in the coming years. With emerging markets and technological advancements enhancing the experience, avitourism is well-positioned to play an important role in the promotion of sustainable tourism.

An overview of three decades of global research on birdwatching tourists

Birdwatchers, or birders, are recognized as a significant and expanding segment of the ecotourism market, consisting of affluent, well-educated and dedicated individuals (Callaghan *et al.*, 2018). Understanding their profiles, motives and behaviours is important for developing effective management and marketing strategies in the avitourism sector (Li, 2018). Moreover, a deep understanding of avitourism products, tourist demographics and the impacts of tourism activities is essential for promoting sustainability and maximizing economic benefits. The following studies, presented in chronological order, illustrate the evolving understanding of birdwatcher behaviour and preferences, reflecting the diversity within this niche tourism segment. However, it is important to recognize that while these studies provide useful categorizations to understand general trends, birdwatchers do not constitute a homogeneous group. Significant variations in their expertise, motives, preferences and expectations must be considered to avoid overly simplistic classifications.

The exploration begins with McFarlane (1994), who identified four dimensions: affiliative, competitive/achievement, appreciative and the conservation dimension, which encompass learning and contributing to science. Cole and Scott (1999) observed that serious birders exhibit distinct information search behaviours and preferences compared with casual wildlife enthusiasts. As the research evolved, Hvenegaard (2002) introduced a categorization into three levels of specialization among birders: novice, advanced-active and advanced-experienced, noting a positive correlation between the level of specialization and conservation involvement. Scott and Thigpen (2003) discovered four distinct segments of visitors at a birdwatching festival in Texas, each displaying varied behavioural involvement and preferences for settings. Eubanks *et al.* (2004) found significant differences among birder subpopulations across various behavioural, motivational and economic dimensions, indicating a complex picture that challenges simplistic views.

Subsequent studies continued to refine these segments. Connell (2009) argued for recognizing elite birders by their specific skills and knowledge. Kim *et al.* (2010) in Queensland, Australia, identified three clusters of birdwatchers: casual, intermediate and active/committed. In another Australian study, Green and Jones (2010) differentiated 'dedicated birders' who prefer seeking endemic or rare birds; they were also more likely than others to engage in camping and utilize resources beyond travel books or information from birdwatching organizations. Lee *et al.* (2015) discovered that male and female birdwatchers exhibit different birdwatching styles, with females attributing greater importance to personal enrichment, enjoyment, satisfaction and recreation, and males being more likely to perceive the activity as an opportunity to showcase their skills. Conradie and van Zyl (2016) distinguished between casual, active and committed birders, finding notable differences across behavioural indicators.

Werdler (2017) distinguished among three types of birders: (i) 'occasional', who enjoy observing birds both at home and during their holidays; (ii) 'dedicated', who are more likely than others to seek endemic or rare birds, often basing their travel decisions on these preferences; and (iii) 'thrill-seekers', who are primarily motivated by the excitement of spotting spectacular species such as large and colourful birds. Following this, Vas (2017) analysed 200 specialized birdwatching blogs, uncovering the underestimated multidimensionality of avitourism and identifying seven birder subpopulations, including previously overlooked segments such as family and couple birdwatchers. This reveals a richer tapestry of birdwatchers than traditionally recognized, pointing to a diverse community that spans various lifestyles and interests. Similarly, Li (2018) conducted a segmentation study of birdwatchers in Guangzhou, China, proposing three different types of recreation involvement groups with varying levels of environmentally responsible behavioural intentions.

Janeczko *et al.* (2021), in their study in Poland, found that the majority of birdwatchers viewed birdwatching as a permanent hobby rather than a temporary one, considering it more of a lifestyle than a mere pastime; 'being close to nature' and 'fascination with birds' were the two most important reasons for birding. This perspective is supported by Steven *et al.* (2021), who noted that avitourists vary not only in the experiences they seek but also in their dedication to birding, which aligns closely with their expectations and how they perceive birdwatching as part of their identity. Alongside these observations, CBI (2021) provides a structured segmentation of the birdwatching tourism market, distinguishing between casual birders, enthusiastic birders – who both have broader interests beyond birding – and hardcore birders, known for their extensive knowledge and specific preferences, particularly in the birds they seek to observe.

Most recently, Randler and colleagues (Randler, 2023; Randler *et al.*, 2023) identified social, achievement and enjoyment of nature factors as primary motivations for birdwatchers. Advanced German birders, compared with casual and intermediate birders, were found to prefer species with lower body mass and less coloration/brightness, with advanced birders particularly seeking out more challenging species such as nocturnal species, species with larger flight distances, species with smaller population sizes and non-urban birds.

Impacts of Birdwatching Tourism

Sustainable and responsible tourism practices in avitourism, emphasizing conservation and biodiversity protection, are imperative to ensure its long-term sustainability (Schwoerer and Dawson, 2022). This section presents the impacts of birdwatching tourism from a tri-component sustainability lens, i.e. economy, society and environment, as delineated in the relevant literature.

Economic impact

The positive impacts of birdwatching tourism include the increase in regional economic activity of less developed areas, employment opportunities and fundraising for bird conservation efforts. Our current comprehension of the economic value of avitourism largely centres on bird festivals, migration events, renowned birdwatching sites or travel stimulated by the presence of vagrant (birds outside their typical geographic range) and non-vagrant birds (Callaghan *et al.*, 2018). Nicolaides (2013) highlights the potential of avitourism as an economic driver contributing to the sustainable development of South Africa, while also aiding in the preservation of environmental and cultural heritage. This approach ensures the involvement of local communities and other stakeholders in the avitourism growth process. In developing countries, limited economic opportunities often drive local communities to engage in activities that degrade natural resources (Clynes, 2017).

Investment in birdwatching infrastructure and training for local residents can be a sustainable option instead of exploiting natural resources, potentially reducing poverty while preserving the environment (Ocampo-Peñuela and Winton, 2017). With consistent support over time, birdwatching tourism projects have shown to be an effective method for generating employment and improving the monthly earnings of locals who work as birdwatching guides (Biggs *et al.*, 2011). For instance, the unintentional destruction of the Fair Isle Bird Observatory in Scotland in 2019 resulted in a halt to birdwatching tourism, which not only impacted the local economy through lost tourist spending and jobs but also risked the survival of other community services, including air access to the island (Butler, 2021). None the less, it is critical to avoid exaggerating the economic impact of birdwatching tourism, particularly in remote rural and urban areas (Schwoerer and Dawson, 2022). Additionally, the COVID-19 pandemic has highlighted how vulnerable birdwatching tourism can be, with protected areas in developing countries seeing sharp declines in revenue and the number of visitors (Qiao *et al.*, 2023).

Impact on social level and local communities

Birdwatching has a notable impact not only on bird populations but also on human communities. Positive effects on local communities can include enhanced educational opportunities, cultural exchange and the acknowledgement of local expertise in natural history. However, there are also potential adverse effects such as pollution from visitors, damage to habitats, alienation among locals who feel left out, economic benefits leaving the local area and the erosion of culture due to tourism pressures (Sekercioglu, 2002). Projects like those led by *Audubon Magazine* (Clynes, 2017) strive to create local economic benefits and encourage conservation efforts through bird-related tourism. One such initiative in Latin America and the Caribbean has shown encouraging outcomes, including local engagement in tourism training programmes, active local promotion of bird-based tourism and a variety of events like birding festivals that highlight the cultural and natural heritage of the regions (Clynes, 2017).

Impact on avian communities, wildlife and broader environmental level

Previous research has underscored the negative consequences of nature-based recreation on bird populations. Birds are a primary attraction for visitors at birdwatching sites (Guimarães *et al.*, 2015). A clear cyclical pattern has been noted where an increase in bird sightings tends to draw more visitors, which in turn affects the behaviour of birds and their interactions with tourists (Collins-Kreiner *et al.*, 2013). Additionally, changes in the timing of bird migrations may influence birdwatchers' loyalty to certain destinations (Li *et al.*, 2022).

The environmental impact of birdwatching is significant (Sekercioglu, 2002). Although considered environmentally friendly, birdwatching can adversely affect not only the observed birds but also other species in the area (Green and Jones, 2010). These impacts might include physiological changes in birds, alterations in their behaviour, variations in population and effects on their breeding success (Steven *et al.*, 2011). For example, tourist traffic could influence where birds choose to nest, potentially endangering their breeding outcomes. In a study examining bird responses to regular boat traffic during birdwatching tours in Spain, McFadden *et al.* (2017) observed that disturbances were more frequent, and more birds were affected, in narrow tidal channels where boats pass closer to the birds.

Beyond the immediate impacts on birds, Kronenberg (2014) explored how birdwatching affects the use of ecosystem services, categorizing four types of indirect impacts: (i) those that alter the ecosystem's ability to provide other services; (ii) those affecting other ecosystem elements; (iii) additional consumption of products, infrastructure or services needed to utilize a specific ecosystem service and their environmental impacts over their life cycle; and (iv) broader societal impacts.

The environmental cost of travel associated with twitching also warrants serious consideration, as it often involves extensive, frequent trips across the country, which have their own negative externalities (Kronenberg, 2014; Brock *et al.*, 2021). It is also crucial to acknowledge that birds are exceptionally vulnerable to climate change impacts, which can alter their distribution, migration timings, population size, community composition and physical characteristics (Underhill and Barnard, 2009).

A Stakeholder Perspective of Sustainable Birdwatching Tourism

This section delves into the sustainability practices within birdwatching tourism by examining a range of stakeholders in the UK, where birdwatching has deep historical roots and significant market presence. In fact, about three million people engage in birdwatching every year in the UK (Tapper, 2022). Our investigation involves a preliminary and exploratory analysis of online content from diverse stakeholders, including tour operators, a destination management organization (DMO), a specialized sustainable birdwatching company and media sources. The aim is to explore how these entities integrate sustainability into their operational and promotional strategies to provide insights into current market practices. The selection is, of course, not representative nor exhaustive of the entire market but is chosen to illustrate a broad spectrum of sustainability initiatives. As such, it serves as a useful starting point to complement the review of extant literature with market perspectives that may highlight emerging trends or underscore overlooked issues. Table 6.1 provides an overview of how key stakeholders integrate sustainability into their operations, illustrating specific content, initiatives and the impacts these have on promoting sustainability in the birdwatching tourism sector.

Each entry in Table 6.1 outlines specific initiatives and their impacts, demonstrating diverse commitments not only to preserving bird habitats but also to enhancing local communities and reducing the ecological footprint of tourism. This preliminary analysis shows a broad commitment across various entities to integrate sustainability into their business models. However, the depth and focus of these sustainability efforts vary among the stakeholders. While some place a pronounced emphasis on community and economic development, others prioritize direct environmental impacts, implementing practices that protect ecosystems and biodiversity. In this context, educational resources also play a crucial role in this spectrum by disseminating knowledge and raising awareness about the importance of sustainable practices and conservation. This range of approaches underscores the complex interplay between economic, social and environmental goals within the birdwatching tourism sector. At the same time, these varied emphases highlight the ongoing challenge within the industry to harmonize economic development with environmental stewardship.

Discussion

This chapter concludes by drawing together the critical sustainability challenges identified in birdwatching tourism. Our discussion is rooted in an examination of both scholarly research and practical insights from key industry stakeholders. The following subsection presents these challenges, providing a foundation for exploring strategic solutions in the subsequent discussion.

Birdwatching tourism: sustainability challenges

The insights drawn from our literature review and analysis of birdwatching stakeholders' websites shed light on sustainability challenges inherent in birdwatching tourism. Based on our investigation, both academics and market stakeholders recognize the complexity of these challenges; however, they often seem to emphasize different aspects. The academic focus primarily revolves around the environmental and economic impacts of birdwatching tourism. These academic considerations delve into the conservation issues, the economic benefits and pressures of birdwatching on local ecosystems, as well as on the broader implications of these activities on sustainable tourism practices.

On the other hand, stakeholders appear to emphasize the potential of birdwatching to

Table 6.1. Birdwatching sustainability initiatives across stakeholders.

Source	Sustainability content	Key initiatives	Impact on sustainability
Birda community app (https://birda.org/, accessed 15 October 2024) A mobile application designed to connect birdwatchers with nature through interactive features and community engagement	Facilitates deep engagement with nature through digital guides and challenges. Promotes community-led conservation efforts	*Community empowerment*: Uses technology to educate on bird conservation *Active conservation*: Engages users in data collection and species tracking	Enhances conservation knowledge and active participation in environmental stewardship
Nature Trek (https://www.naturetrek.co.uk/, accessed 15 October 2024) A travel company that organizes birdwatching tours globally, focusing on natural experiences	Offers environmentally sensitive travel options, including tours that minimize travel emissions. Employs local guides to enhance the visitor experience and benefit the community	*Eco-friendly travel solutions*: Advocates for slow and no-fly travel options *Local employment*: Prioritizes hiring local guides to boost local economy	Reduces carbon footprint and supports local economic growth through sustainable tourism practices
The Sustainable Birding Company (https://sustainablebirding.com/about/, accessed 15 October 2024) A company dedicated to providing birdwatching experiences that emphasize conservation and responsible tourism	Committed to bird species' conservation, carbon footprint offsetting and enhancing local community welfare through sustainable tourism	*Conservation initiatives*: Implements projects to protect bird habitats *Community support*: Engages in programmes that benefit local populations economically	Leads in sustainable practices by balancing ecological goals with community development
Responsible Travel (https://www.responsibletravel.com/, accessed 15 October 2024) A travel agency that curates environmentally conscious travel experiences across various destinations	Promotes accessible and responsible tourism packages that are designed to have minimal environmental impact and support local communities	*Responsible packages*: Offers flight-free travel and volunteer opportunities *Community integration*: Ensures that tourism benefits are shared with local communities	Advocates for a holistic approach to responsible tourism that includes environmental, economic and social dimensions

Continued

Table 6.1. Continued

Source	Sustainability content	Key initiatives	Impact on sustainability
Visit Norfolk (DMO), UK (https://www.visitnorfolk.co.uk/post/birdwatching-in-norfolk, accessed 15 October 2024) The official tourism board for Norfolk, UK, promoting local attractions including wildlife and birdwatching spots	Provides educational resources at eco-friendly visitor centres and promotes birdwatching as a means to learn about and protect avian life	*Education and outreach*: Focuses on enhancing visitor knowledge and appreciation for local biodiversity *Eco-centric facilities*: Develops visitor centres that use sustainable materials and practices	Promotes environmental education and sustainability in tourism infrastructure
British Birds Journal (https://britishbirds.co.uk/journal, accessed 15 October 2024) A journal blending rigorous ornithological research with articles for enthusiastic birdwatchers, focusing on bird conservation and observation	Advocates for bird conservation through scholarly articles, promoting ethical birdwatching practices and habitat protection	*Educational content*: Provides in-depth analysis on bird species and conservation issues *Community engagement*: Fosters a knowledgeable community of birdwatchers	Increases awareness and knowledge among birdwatchers, contributing to conservation-minded practices
Bird Watching Magazine (https://www.birdwatching.co.uk, accessed 15 October 2024) A magazine offering insights, news and tips for birdwatchers, with a focus on bird conservation and birdwatching techniques	Features articles on sustainable birdwatching, species protection and environmental impacts of birdwatching	*Conservation awareness*: Highlights initiatives and challenges in bird conservation *Sustainability pledges*: Commits to educating birdwatchers on reducing their environmental impact	Enhances understanding of bird conservation issues and promotes sustainable practices within the birdwatching community

engage people with nature and promote environmental stewardship. Market perspectives also point to the social and educational benefits of birdwatching. At the same time though, not all stakeholders consistently highlight the importance of conserving nature, respecting local communities and minimizing adverse environmental effects. These insights are of course drawn from a preliminary examination of a selection of stakeholders. However, they do serve to indicate that while some stakeholders prioritize and articulate their commitment to sustainable practices on their platforms, others tend to emphasize the commercial or experiential aspects of birdwatching tourism, without explicitly addressing broader sustainability issues.

One major sustainability challenge identified through our investigation is the disturbance to wildlife and habitats caused by birdwatching tourism activities. The literature and stakeholder content reveal that increased traffic in bird habitats can lead to habitat degradation and behavioural changes in birds, potentially threatening their survival and biodiversity. This is a critical concern as it highlights a tension between the desire to observe and appreciate wildlife and the unintended consequences of doing so. In this connection, one needs to remember that while birdwatching tourism may bring economic benefits to local economies and conservation projects, this relationship becomes unsustainable if the viability of local ecosystems is jeopardized by wildlife disturbance (Lindsay *et al.*, 2008).

Another significant challenge discussed is the carbon footprint associated with travel to birdwatching sites. This concern is particularly pronounced in the context of twitching: how can a tourism activity be considered sustainable when, for example, travelling to a long-haul destination entails a high carbon footprint? Stakeholder websites and academic studies both point to the paradox of this practice. While birdwatching is intended to connect people with nature and foster a sense of conservation, the travel involved can contribute substantially to carbon emissions, thus undermining the activity's environmental goals. Birdwatchers in the UK who are mindful of their environmental impact are promoting low-carbon birding practices by focusing on local areas instead of travelling long distances (Tapper, 2022).

Economic benefits from birdwatching tourism are widely acknowledged but are accompanied by their own set of challenges. Indeed, for economic benefits to be realized, increased development in remote areas is often required. However, this development can threaten the very natural environments that attract birdwatchers if the development is not managed sustainably (Collins-Kreiner *et al.*, 2013). In this connection, the analysis of stakeholder websites indicates a recognition of the need to develop birdwatching tourism responsibly, ensuring that it supports both the local economy and the conservation of natural resources. Social impacts are also an area of concern, particularly the potential for local communities to feel left out from tourism benefits (Sekercioglu, 2002). Insights from stakeholders, which predominantly highlight sustainability initiatives and community involvement in birdwatching tourism, are relevant in this respect and demonstrate a commitment to sustainability and inclusion.

Addressing sustainability challenges in birdwatching tourism

Addressing the sustainability challenges identified in the previous subsection requires a multifaceted approach. Collins-Kreiner *et al.* (2013) advocate for the Limits of Acceptable Change (LAC) framework as a strategy to ensure sustainability in birdwatching tourism. This framework, which focuses on measuring the impacts of tourism, establishing distinct zones for various activities and tailoring the LAC to suit each location and season, is precisely aligned with the sensitivity of both the tourism activities and bird populations involved. As such, it addresses concerns about the impact of tourism on sensitive bird habitats and populations, managing these impacts through zoning and tailored management strategies.

In addressing the broader sustainability challenges of birdwatching tourism, such as the impacts on wildlife habitats and the balance of economic benefits, regions with unique geographic or sociopolitical characteristics are often faced with additional challenges. These include navigating land-use conflicts and ownership disputes that may arise as tourism develops. Such conflicts can threaten the conservation goals if not managed carefully, necessitating targeted protective measures. For example, enforcing regulations to restrict boat traffic in sensitive birdwatching areas helps maintain crucial buffer zones that protect wildlife from human disturbances (McFadden *et al.*, 2017), thus aligning with the conservation efforts outlined earlier. Moreover, the effectiveness of these interventions often depends on robust local governance, adequate infrastructure and the active engagement of local communities in both planning and implementing sustainable practices (Werdler, 2017).

Furthermore, in keeping with the issue of community engagement, integrating local communities in birdwatching tourism is vital for its success and sustainability. This strategy not only supports economic opportunities but also fosters biodiversity conservation and the preservation of environmental and cultural heritage (Nicolaides, 2013). Establishing a symbiotic relationship between tourism and conservation ensures that local support for birdwatching initiatives contributes effectively to the conservation of both native and migratory bird species, thus creating strong connections between local stakeholders and residents (Ramos *et al.*, 2023).

On a broader scale, mitigating the carbon footprint associated with travel to birdwatching sites addresses the paradox of promoting conservation-oriented tourism while managing its environmental impacts. Although ecological birdwatching may incur higher costs per capita than general ecotourism, it yields a greater marginal value and provides a broader range of social, economic and environmental benefits, showcasing its potential as a sustainable tourism model with appropriate management and cooperation among stakeholders (Liu *et al.*, 2021). Adhering to ethical birdwatching guidelines and supporting local businesses that engage in low-impact practices are also essential. The American Birding Association's mission (ABA, 2024), for example, encourages birdwatchers to use their skills in conserving bird habitats and promoting ethical birdwatching practices, which include a specific code of ethics that emphasizes the welfare of birds and the integrity of birdwatching activities. Moreover, mitigating negative behaviours among birdwatchers through self-regulation and community monitoring is an effective grassroots approach to enforcing responsible behaviour (Callaghan *et al.*, 2020).

Ensuring sustainability in birdwatching tourism also involves environmental education, enforcement of conservation laws and the adoption of new technologies. Environmental and avitourism literacy, as emphasized by Conradie and van Zyl (2020), is critical for raising awareness and promoting responsible behaviours among tourists and locals. Efforts should therefore be directed towards developing comprehensive educational programmes that encompass not only the basics of birdwatching but also the ethical implications and conservation strategies involved. Educational initiatives like BirdBot (https://www.bird.bot, accessed 15 October 2024) are important in this respect, advocating for eco-friendly birdwatching practices, including responsible travel, habitat conservation and sustainable bird feeding.

Furthermore, the enforcement of conservation laws must be strengthened to protect sensitive habitats and species from the pressures of tourism. This involves not only the application of existing regulations but also the development of new policies that address emerging environmental challenges. Technological innovations also play a crucial role in enhancing sustainability in birdwatching tourism. Tools such as TerraPass (https://terrapass.com, accessed 15 October 2024) for carbon offsetting and eBird (https://ebird.org, accessed 15 October 2024) for tracking bird species sightings are instrumental in these efforts.

Lastly, enhancing the marketing and communication strategies of birdwatching activities to better reflect sustainability goals can significantly contribute to the sector's overall sustainability. This includes developing strategic marketing plans that encompass sustainable product options, pricing strategies to support conservation efforts and sustainability-focused promotional campaigns. Establishing strategic alliances that focus on sustainability throughout the tourism value distribution chain is imperative for the design, provision and promotion of sustainable tourism offerings (Collins-Kreiner *et al.*, 2013).

Conclusions

Birdwatching tourism, as explored in this chapter, presents an opportunity to align economic, social and environmental sustainability. However, the sustainability challenges identified require comprehensive strategies to ensure that birdwatching tourism remains a positive influence on both nature and communities. At the same time, successfully addressing these challenges could create a more sustainable and enjoyable environment for birdwatchers, which in turn may enhance their overall experience.

To fully realize the benefits of birdwatching tourism, however, it is crucial to adopt an integrative approach that goes beyond addressing immediate challenges. Towards this direction, stakeholders must work together to create a resilient tourism model that anticipates future environmental and economic shifts and developments. This involves not only mitigating current impacts but also strengthening the ability of both human societies and natural environments to adapt and respond to changing conditions. For example, fostering a culture of continuous learning and adaptation within local communities can empower them to manage and benefit from tourism in a responsible manner. Moreover, integrating advanced technologies and innovative practices can help monitor environmental impacts more effectively and improve the overall quality of the birdwatching experience. Such efforts should be supported by strong policy frameworks that encourage sustainable practices among all stakeholders and ensure compliance with conservation standards.

In summary, realizing the potential of birdwatching tourism requires confronting its inherent sustainability challenges. By implementing strategies in line with those identified in this chapter, birdwatching tourism can address these challenges and progress towards sustainable growth. This approach aims to ensure that all stakeholders derive practical benefits from this growing industry while actively contributing to its sustainability, moving beyond theoretical discussions to establish a realistic and sustainable future for birdwatching tourism.

References

ABA (2024) American Birding Association homepage. Available at: https://www.aba.org/american-birding-association/ (accessed 15 October 2024).

Biggs, D., Turpie, J., Fabricius, C. and Spenceley, A. (2011) The value of avitourism for conservation and job creation – an analysis from South Africa. *Conservation and Society* 9(1), 80–90.

Brock, M., Fraser, I., Law, C., Mitchell, S. and Roberts, D.L. (2021) An economic analysis of twitching behaviour and species rarity. *Journal of Environmental Economics and Policy* 10(1), 54–73.

Butler, R.W. (2021) The potential impacts of climate change on avitourism to islands: The example of Fair Isle, Scotland. *International Journal of Islands Research* 2(1), 4. DOI: 10.21427/683v-tc36.

Callaghan, C.T., Slater, M., Major, R.E., Morrison, M., Martin, J.M. *et al.* (2018) Travelling birds generate eco-travellers: The economic potential of vagrant birdwatching. *Human Dimensions of Wildlife* 23(1), 71–82.

Callaghan, C.T., Benson, I., Major, R.E., Martin, J.M., Longden, T. *et al.* (2020) Birds are valuable: The case of vagrants. *Journal of Ecotourism* 19(1), 82–92.

CBI (2021) The European market potential for birdwatching tourism. Available at: https://www.cbi.eu/market-information/tourism/birdwatching-tourism/market-potential (accessed 15 October 2024).

Clynes, T. (2017) How birding tourism can support communities and foster conservation. National Audubon Society. Available at: https://www.audubon.org/news/how-birding-tourism-can-support-communities-and-foster-conservation (accessed 5 April 2024).

Cole, J.S. and Scott, D. (1999) Segmenting participation in wildlife watching: A comparison of casual wildlife watchers and serious birders. *Human Dimensions of Wildlife* 4(4), 44–61.

Collins-Kreiner, N., Malkinson, D., Labinger, Z. and Shtainvarz, R. (2013) Are birders good for birds? Bird conservation through tourism management in the Hula Valley, Israel. *Tourism Management* 38, 31–42.

Connell, J. (2009) Birdwatching, twitching and tourism: Towards an Australian perspective. *Australian Geographer* 40(2), 203–217.

Conradie, N. and van Zyl, C. (2016) Behavioural involvement in avitourism: An international case study. *African Journal of Hospitality, Tourism, and Leisure* 5(2), 1–23.

Conradie, N. and van Zyl, C. (2020) A conceptual literacy framework for sustainable avitourism to ensure quality birding experiences. In: Gunadi, I.M.A., Kausar, D.R., Firmansyah, R. and Kennell, J. (eds) *Culture, People and Technology – The Driving Forces for Tourism Cities: Proceedings of 8th ITSA Biennial Conference 2020*, International Tourism Studies Association, p. 586.

Eubanks, T.L. Jr, Stoll, J.R. and Ditton, R.B. (2004) Understanding the diversity of eight birder sub-populations: Socio-demographic characteristics, motivations, expenditures and net benefits. *Journal of Ecotourism* 3(3), 151–172.

Future Market Insights (2024) Ornithology Tourism Market Outlook from 2024 to 2034. Available at: https://www.futuremarketinsights.com/reports/ornithology-tourism-sector-perspective-and-outlook (accessed 15 October 2024).

Gosálvez Rey, R.U., Navas Berbel, A. and Aleja, D. (2022) Birdwatching as a new tourist activity in El Hierro Geopark. In: Dóniz-Páez, J. and Pérez, N.M. (eds) *El Hierro Island Global Geopark: Diversity of Volcanic Heritage for Geotourism*. Springer, Cham, Switzerland, pp. 95–104.

Green, R. and Jones, D.N. (2010) *Practices, Needs and Attitudes of Bird-Watching Tourists in Australia*. CRC for Sustainable Tourism Pty Ltd, Gold Coast, Australia.

Grihault, N. (2023) 8 of the best birdwatching destinations around the world. Available at: https://restless.co.uk/travel/the-best-birdwatching-destinations-around-the-world/ (accessed 10 April 2024).

Guimarães, M.H., Nunes, L.C., Madureira, L., Santos, J.L., Boski, T. *et al*. (2015) Measuring birdwatchers preferences: A case for using online networks and mixed-mode surveys. *Tourism Management* 46, 102–113. DOI: 10.1016/j.tourman.2014.06.016.

Hvenegaard, G.T. (2002) Birder specialization differences in conservation involvement, demographics, and motivations. *Human Dimensions of Wildlife* 7(1), 21–36.

Janeczko, E., Łukowski, A., Bielinis, E., Woźnicka, M., Janeczko, K. *et al*. (2021) "Not just a hobby, but a lifestyle": Characteristics, preferences and self-perception of individuals with different levels of involvement in birdwatching. *PLoS One* 16(7), e0255359.

Kim, A.K., Keuning, J., Robertson, J. and Kleindorfer, S. (2010) Understanding the birdwatching tourism market in Queensland, Australia. *Anatolia* 21(2), 227–247.

Kronenberg, J. (2014) Environmental impacts of the use of ecosystem services: Case study of birdwatching. *Environmental Management* 54(3), 617–630.

Lee, C.K., Lee, J.H., Kim, T.K. and Mjelde, J.W. (2010) Preferences and willingness to pay for birdwatching tour and interpretive services using a choice experiment. *Journal of Sustainable Tourism* 18(5), 695–708.

Lee, S., McMahan, K. and Scott, D. (2015) The gendered nature of serious birdwatching. *Human Dimensions of Wildlife* 20(1), 47–64.

Li, L. (2018) Environmentally responsible behavior intentions of wildlife tourists based on recreational involvement: A case study of bird-watching tourists in Guangzhou. *Chinese Studies* 7(1), 95–113.

Li, N., Yang, C., Qin, X., Sun, J. and Liu, J. (2022) The impact of climate change on birder destination loyalty: Examining changes in bird resources. *Current Issues in Tourism* 25(11), 1798–1816.

Lindsay, K., Craig, J. and Low, M. (2008) Tourism and conservation: The effects of track proximity on avian reproductive success and nest selection in an open sanctuary. *Tourism Management* 29(4), 730–739.

Liu, T., Ma, L., Cheng, L., Hou, Y. and Wen, Y. (2021) Is ecological birdwatching tourism a more effective way to transform the value of ecosystem services? A case study of birdwatching destinations in Mingxi county, China. *International Journal of Environmental Research and Public Health* 18(23), 12424.

MarkWide Research (2024) Ornithology Tourism Market Analysis – Industry Size, Share, Research Report, Shares, Covid-19 impact, Statistics, Trends, Growth and Forecast 2024–2032. Available at: https://markwideresearch.com/ornithology-tourism-market/ (accessed 3 April 2024).

McFadden, T.N., Herrera, A.G. and Navedo, J.G. (2017) Waterbird responses to regular passage of a birdwatching tour boat: Implications for wetland management. *Journal for Nature Conservation* 40, 42–48.

McFarlane, B.L. (1994) Specialization and motivations of birdwatchers. *Wildlife Society Bulletin* 22(3), 361–370. Available at: https://www.jstor.org/stable/3783377 (accessed 15 October 2024).
Nicolaides, A. (2013) Promoting avitourism as a special niche area of ecotourism in South Africa. *African Journal of Hospitality, Tourism and Leisure* 2(3), 1–16.
Ocampo-Peñuela, N. and Winton, R.S. (2017) Economic and conservation potential of bird-watching tourism in postconflict Colombia. *Tropical Conservation Science* 10, 1940082917733862.
Qiao, H., Orr, M., Yang, Q., Zhan, X., Lei, F. *et al.* (2023) Global birdwatching data reveal uneven consequences of the COVID-19 pandemic. *Biological Conservation* 288, 110351.
Ramos, K., Pérez, L.R. and Bassols, N. (2023) Bird watching tourism in sustainable and accessible destinations: The case of the entrepreneurial venture Explora Caloccita in the Mexican Pacific Coast. In: Allis, T. and Carnicelli, S. (eds) *Tourism in Latin America: Current Practices and Future Developments*. Tourism Cases Special Issue. CAB International, Wallingford, UK. DOI: 10.1079/tourism.2023.003.
Randler, C. (2023) Motivations for birdwatching: Support for a three-dimensional model. *Human Dimensions of Wildlife* 28(1), 84–92.
Randler, C., Staller, N., Kalb, N. and Tryjanowski, P. (2023) Charismatic species and birdwatching: Advanced birders prefer small, shy, dull, and rare species. *Anthrozoös* 36(3), 427–445.
Ren, J., Su, K., Zhou, Y., Hou, Y. and Wen, Y. (2022) Why return? Birdwatching tourists' revisit intentions based on structural equation modelling. *Sustainability* 14(21), 14632.
Riley, A. (2014) South Africa: The best of the 'birding continent'. In: *Birder's Guide*, Vol. 26. Available at: https://www.aba.org/south-africa-best-birding-continent/ (accessed 10 April 2024).
Sánchez-Rivero, M., Sánchez-Martín, J.M. and Rodríguez Rangel, M.C. (2020) Characterization of birdwatching demand using a logit approach: Comparative analysis of source markets (national vs foreign). *Animals* 10(6), 965.
Schwoerer, T. and Dawson, N.G. (2022) Small sight-big might: Economic impact of bird tourism shows opportunities for rural communities and biodiversity conservation. *PLoS One* 17(7), e0268594.
Scott, D. and Thigpen, J. (2003) Understanding the birder as tourist: Segmenting visitors to the Texas hummer/bird celebration. *Human Dimensions of Wildlife* 8(3), 199–218.
Sekercioglu, C.H. (2002) Impacts of birdwatching on human and avian communities. *Environmental Conservation* 29(3), 282–289.
Stemmer (formerly Jathe), K., Aas, Ø., Veisten, K. and Lindberg, K. (2022) Assessing recreation specialization to guide nature-based tourism development: A hybrid choice model of birder destination preferences. *Journal of Outdoor Recreation and Tourism* 39, 100516.
Steven, R., Pickering, C. and Castley, J.G. (2011) A review of the impacts of nature based recreation on birds. *Journal of Environmental Management* 92(10), 2287–2294.
Steven, R., Morrison, C. and Castley, J.G. (2015) Birdwatching and avitourism: A global review of research into its participant markets, distribution and impacts, highlighting future research priorities to inform sustainable avitourism management. *Journal of Sustainable Tourism* 23(8–9), 1257–1276.
Steven, R., Rakotopare, N. and Newsome, D. (2021) Avitourism tribes: As diverse as the birds they watch. In: Pforr, C., Dowling, R. and Volgger, M. (eds) *Consumer Tribes in Tourism*. Springer, Singapore, pp. 101–118. DOI: 10.1007/978-981-15-7150-3_8.
Tapper, J. (2022) You don't need to travel long distances to spot birds, Britain's twitchers urged. Available at: https://www.theguardian.com/environment/2022/jan/01/travel-distances-britain-twitchers-birds-birdwatchers-species-foot-cycle (accessed 15 October 2024).
Underhill, L. and Barnard, P. (2009) SABAP2 and climate change. *Africa – Birds & Birding* 14(6), 79. Available at: https://science.uct.ac.za/sites/default/files/content_migration/science_uct_ac_za/925/files/ABB14%25286%252978-79.pdf (accessed 15 October 2024).
Vas, K. (2017) Birding blogs as indicators of birdwatcher characteristics and trip preferences: Implications for birding destination planning and development. *Journal of Destination Marketing & Management* 6(1), 33–45.
Werdler, K. (2017) Birdwatching tourism sector in Rwanda: The need of self-learning and self-improving for the future of wildlife tourism development. In: Green, R.J. (ed.) *Wildlife Tourism, Environmental Learning and Ethical Encounters. Geoheritage, Geoparks and Geotourism*. Springer, Cham, Switzerland, pp. 43–56.
Xu, J., Yan, P., Liu, Z., Qin, H. and Jiang, A. (2023) Avitourism as an aspect of sustainable mountain development: A case study from Southern China. *Eco.Mont (Journal on Protected Mountain Areas Research)* 15(1), 38–45.

7 Sustainability in the Wine Tourism Industry: Perceptions, Practices and Challenges of Family Businesses

Stella Kladou[1], Maria Alebaki[2]* and Kleio Arvanakou[3]

[1]Department of Business Administration and Tourism, Hellenic Mediterranean University, Crete, Greece; [2]Agricultural Economics Research Institute, Hellenic Agricultural Organization DIMITRA, Greece; [3]School of Social Sciences, Hellenic Open University, Patras, Greece

Abstract

Wine tourism has emerged as an example of coexistence and co-support between the wine and tourism industries. Despite the progress made during the last few decades and the great potential of the sector to provide sustainable solutions for both wineries and destinations, limited literature is available on how such goals may be achieved. This chapter draws from an ongoing project which explores sustainability practices and challenges for family wineries in an Old-World context. To address this objective, we collected data from island and mainland destinations in Greece. The current chapter employed the Gioia *et al.* (2013) methodology to analyse empirical results from the island of Corfu. Content analysis of the winery websites and additional in-depth interviews and a census survey with wine tourism stakeholders helped apply Teece's *et al.* (1997) model of dynamic capabilities in the case of pursuing sustainable wine tourism. Analysis reveals the importance of not only sensing but also of seizing and configuring opportunities and challenges in order to achieve sustainable wine tourism development. Along these lines, a holistic understanding of sustainability, adequate education and training, as well as effective collaborations within and beyond the wine industry and wine tourism sector emerge as crucial.

Keywords: Family winery, sustainable wine tourism, collaboration, resilience, Greece

Introduction

The development of alternative tourism forms and niche tourism emerges as a necessity in the pursuit of sustainability. Yet, niche tourism types may not, by definition, align with sustainability standards across all sustainability pillars (i.e. environmental, sociocultural, economic). Specifically in the case of wine tourism, Nave *et al.* (2021) revealed supply- and demand-related factors explaining the adoption of sustainable practices. In relation to the former, competitiveness, innovation and territorial development relate to practices reducing the ecological footprint and the adaptation to new forms of consumption which boost economic and social development. In relation to the latter, wine consumers become increasingly more aware of sustainability, choose responsible lifestyles and thus prefer wineries that adopt sustainable practices.

*Corresponding author: malempaki@gmail.com

DOI: 10.1079/9781800626669.0007

Wine businesses, in response, may generally prioritize sustainability goals and objectives (Nave *et al.*, 2021). Yet, it remains interesting to investigate the link between sustainability and wine tourism development through the lens of the dynamic capabilities approach (Teece *et al.*, 1997).

This chapter focuses on a wine destination currently at the developing stage of the wine tourism life cycle, namely the island of Corfu in Greece. The chapter describes a study that explored wine producers' understanding of sustainability along with their sustainability efforts, given the opportunities and challenges which they encounter. A mixed methodology, pairing data collected through a structured questionnaire with information provided through semi-structured interviews and participant observation, facilitated a deep understanding of the relationship between wine tourism and sustainability as experienced and developed by local wineries.

Inductive analysis of the findings, through the use of Gioia methodology (Gioia *et al.*, 2013) and the interpretation of the dynamic capabilities approach, demonstrates how wine tourism has started developing at the destination. Findings suggest that stakeholders are familiar with the concept. Yet, their understanding of sustainability is quite limited and usually refrains itself to environmental sustainability. The study puts forward a theoretical framework that facilitates involved stakeholders' understanding and, thus, helps them harvest the benefits of sustainable wine tourism development. This framework reveals that what matters the most for the latter is not the number of wineries involved in such initiatives but the holistic understanding of sustainability of those wineries involved in tourism activities. Moreover, such sustainable initiatives are possible only once winemakers are adequately trained and educated on sustainable practices, learn how to become part of relevant planning and collaborate towards one common goal (i.e. sustainable wine tourism development). Still, the latter is possible when other local products of high quality are also found at the destination, thereby suggesting the need for collaboration with companies and stakeholders beyond the wine sector.

Empirical findings of this work provide practitioners with guidelines on the importance of the theory of dynamic capabilities (i.e. sensing, seizing and reconfiguring the opportunities in order to achieve sustainable wine tourism development). Finally, wine and sustainable tourism scholars once again come to realize the complexity and interdisciplinary nature of sustainable wine tourism research, while the study paves the ground for future research that will help understand how to build win–win sustainable alliances between the wine and the tourism industry.

Theoretical Background

Wine tourism and sustainability

Niche tourism as a concept is not new; it developed as a response to travellers' motives and decision-making patterns which favour travellers' interest in performing specific activities while travelling or in visiting specific destinations (Hall and Weiler, 1992). Along these lines, wine tourism developed as a multifaceted phenomenon which combines all those tangible and intangible attributes that make up the wine experience. For instance, Hall and Mitchell (2000, p. 447) define wine tourism as 'visitation to vineyards, wineries, wine festivals and wine shows for which grape wine tasting and/or experiencing the attributes of a grape wine region are the prime motivating factors for visitors'.

Wine tourism is not by definition nor 'from birth' intertwined with sustainability. Yet, sustainability emerges as a priority for both the wine and tourism industries in order to correspond to increasing consumption and demand at a global level. Thus, scholars investigated the sustainability of wine regions in an effort to develop a sustainable model for wine tourism, which needs to incorporate sociocultural, environmental and economic factors (e.g. Poitras and Getz, 2006; Montella, 2017; Serra-Cantallops *et al.*, 2021). According to Stasi *et al.* (2016), we can recognize three sets of factors which make it essential to prioritize sustainable wine tourism development nowadays. These factors are interlinked, favour the sustainable development of both the winery and its corresponding location, and concern the following:

1. *External factors.* Wine lovers and consumers are increasingly more conscious, prefer certified products even if these come at a higher price, and seek innovative packages which combine leisure, wine tasting and wine experiences that reflect sustainable practices (Forbes *et al.*, 2009; Flores, 2018). Local communities may also recognize sustainable practices as a priority and seek to protect their ecosystem in order to build a competitive advantage (Montella, 2017; Serra-Cantallops *et al.*, 2021). The strategic collaboration and co-creation of value between the local communities and wine industry stakeholders builds on sustainable tourism models and suggests that winemakers need to acknowledge local resources and share common ethical and environmental values (Lane, 1994; Williams and Dossa, 2003). External factors also include governmental policies which seek to develop and support sustainable practice and develop consciousness.
2. *Internal factors* relate to winemakers' own concerns relevant to the environmental pillar of sustainability (e.g. waste management, water supply/shortage) (Getz and Brown, 2006) and their social responsibility as employers (Flores, 2018), balancing profit with the carrying capacity of the region (Williams and Dossa, 2003).
3. *Strategic factors* relate to image and the communication mix of the wineries. Wineries that respect the environment and adopt sustainable practices can build a competitive advantage on these attributes and attract loyal customers. Besides, their sustainable products support feelings of trust and safety and, in extent, raise profits (Mastroberardino *et al.*, 2020).

In sum, sustainable wine tourism relates to both supply- and demand-related perspectives. As Nave *et al.*'s (2021) review reveals, supply-side aspects reflect wineries which see sustainability practices as factors boosting their competitiveness, adopt relevant innovations in order to mainly reduce their ecological footprint, adapt to new forms of consumption and pursue territorial development in the search for economic and social development. Still, demand-related factors also influence sustainable wine tourism development in response to consumers' increased awareness of and preference for sustainable lifestyles and companies (including wineries) that implement resilient and sustainable practices (Nave *et al.*, 2021). Despite the growing number of studies investigating resilience and sustainability in the wine industry and wine tourism, we argue that wineries and wine tourism stakeholders may be more or less equipped to adopt sustainable practices and innovations based on their own dynamic capabilities. Thus, an exploration of how the theory of dynamic capabilities may apply in wine tourism emerges as necessary (Alonso *et al.*, 2020).

Dynamic capabilities: the theory and its application in wine tourism

Teece *et al.* (1997) were among the first researchers to define dynamic capabilities as the ability to complete/combine and redevelop/transform internal and external resources and acquire new resources. The authors highlight the importance of strategic management with their reference to (cap)abilities. In addition, Barreto (2010) emphasizes the importance of a minimum level of efficiency and the importance of repetition.

According to Teece *et al.* (1997), dynamic capabilities build on: (i) organizational learning; (ii) the alignment of corporate resources and capabilities; and (iii) the transformation of resources and capabilities. More recently, Teece (2007) identified dynamic capabilities in three interlinked stages, which include the following:

1. Stage 1 involves the corporate ability to detect opportunities and threats by *sensing* its internal and external environment. In wine management and tourism in specific, winemakers sense possible opportunities and threats before proceeding to investments (McMullen and Shepherd, 2006).
2. Stage 2 concerns the corporate ability to *seize* the opportunities. This stage might maximize the harvesting of opportunities through appropriate activities and operations in wine tourism and management as well (McMullen and Shepherd, 2006),

for example as reflected on innovating or 'simply' adjusting winery settings in order to facilitate visitation.

3. Stage 3 refers to the ability of the company to maintain its competitive advantage and *reconfigure* its tangible and intangible resources in order to face any threats. Teece (2012, 2014) suggests that this is how the company can avoid stagnation and achieve continuous regeneration and renewal. This stage is also relevant to winemakers who need to constantly stay tuned with wine consumers' needs (Thach *et al.*, 2016) and pursue authenticity in the quest for long-term sustainability (Kim and Bonn, 2016).

External and internal factors influence the development of dynamic capabilities. The former category includes institutional, competitive and technological factors, the relationship with other companies, changes in demand and consumer preferences, and the availability of raw materials (Helfat and Peteraf, 2003; Eriksson, 2014). Human capital is probably the most important internal factor, which generally embraces upper managers' involvement, technological, financial and organizational resources, and the learning processes in place (e.g. Eisenhardt and Martin, 2000; Cepeda and Vera, 2007; Pandza and Thorpe, 2009).

Although dynamic capabilities are unique to each business (Teece *et al.*, 1997), scholars emphasize that some characteristics may stand across different companies (e.g. Eisenhardt and Martin, 2000). Dynamic capabilities may be useful and effective based on various parameters, such as the size and type of business and how dynamic the environment is (Helfat and Winter, 2011; Eriksson, 2014). Alonso *et al.* (2020) elaborated on the theory of dynamic capabilities to evaluate the development and sustainability of relevant wine tourism activities, and also revealed that extant studies share some common stages and characteristics (e.g. innovation adds value to a winery and, by extent, to the promotion of wine tourism in the region). The incorporation of new technologies and equipment may boost the competitive advantage in a tangible manner whereas branding, training and knowledge sharing support competitiveness in an intangible manner (Piening and Salge, 2015).

In sum, the importance of the theory of dynamic capabilities for companies in different industries is well established, and some of its applications are also evident in the wine tourism sector. Such applications become increasingly more crucial given the growing necessity of developing sustainable tourism types, especially in the case of well-established leisure destinations which need to become resilient and move towards more sustainable niche tourism types. In response, this chapter draws on a project focusing on the island of Corfu and applies the Gioia *et al.* (2013) methodology in order to shed light to other similar places which may be well established as tourism destinations yet can become more resilient by turning to sustainable wine tourism development.

Methodology

This chapter sets off to investigate the supply side of sustainable wine tourism development, in order to shed light on the carrying capacity and future prospects of a developing wine destination, namely Corfu in Greece. In order to address the research objective, out of 17 wineries located on the island, the identified population involves only those nine wineries that are adequately equipped to accept visitors (Greek Gastronomy Guide, 2021). In total, eight wineries accepted to participate in a census survey and their responses are analysed for the purposes of this chapter.

A mixed methodology was deemed more appropriate in order to reach the study objective and better understand sustainability practices and challenges for family wineries. A preliminary content analysis of winery websites helped identify some attributes and sustainability practices of the wine industry on the island. The quantitative approach employed a self-administered tool built on extant studies (Getz *et al.*, 1999; Alebaki, 2012; Alonso *et al.*, 2019) which included: (i) profiling questions (e.g. legal status of the winery, annual production, number of employees); (ii) questions on winery involvement with wine tourism (when applicable); (iii) questions on certifications and sustainability certifications which the winery may have; and (iv) questions profiling the winery owners/

managers. Once the winery accepted to contribute to the study, one of the researchers emailed the questionnaire and arranged a meeting for a personal, semi-structured interview.

The qualitative approach explains winemakers' behaviour and motives (Foley and Timonen, 2015) and involves interviewing Corfu winemakers on their attitudes towards wine tourism as well as prospects of sustainable wine tourism development on Corfu. Analysis revealed the need to interview two more experts in an effort to thoroughly understand the sector and subsequently propose a theoretical framework of value for developing wine destinations. These experts were a wine critic/judge with international experience and a businessman with in-depth knowledge of the Corfu wine sector. Finally, *in situ* involvement as a wine tourist allowed the researcher to observe the current state of wine tourism on the island.

Semi-structured interviews, with eight winemakers/managers of wineries and two experts, aimed to: (i) assess the current state of wine tourism on the island and identify relevant opportunities and threats; (ii) explore winemakers' goals of wine tourism (when applicable); (iii) address sustainable practices for winemaking and wine tourism on Corfu; (iv) recognize relevant communication activities; (v) discover winemakers' own future plans in terms of wine tourism; and (vi) identify winemakers' perspectives in terms of the future of wine tourism on the island. Each interview lasted approximately 30 min. The interviews took place in Greek, and, in line with research ethics, the interviewees were asked to give their consent to record and analyse their views and also were informed about anonymity and the option to withdraw and refuse to answer any question at any point. Transcription and analysis were undertaken by two researchers.

Data analysis followed the Gioia *et al.* (2013) methodology and Teece *et al.* (1997) afore-described model of dynamic capabilities. In specific, Gioia *et al.* (2013) suggested to first isolate responses relevant to opportunities and challenges which feed into motives (i.e. similarities and differences). Then, first-order themes feed into motives (i.e. similarities and differences) and are subsequently organized as second-order themes. These second-order themes are then interpreted as conceptual phrases/concepts with the aim to create a unified theoretical dimension. According to Gioia *et al.* (2013), these three stages allow for validity in qualitative research.

Wine Tourism on Corfu: An Overview

Greece is among the most popular destinations for leisure, where wine tourism development started mainly in the 1990s (Alebaki and Iakovidou, 2010). One of the first references to wine on Corfu is attributed to Homer in *Odyssey*, thereby confirming the historic and cultural value of the sector for the island (Corfu Food and Wine Festival, 2018). Vineyards expanded towards the end of the 14th century until the Venetian rule of the island limited wine production to support the production of olive oil instead (Corfu Food and Wine Festival, 2018). Napoleon and subsequent British rule supported wine production until the 1960s, when the local population started migrating to urban centres and Corfu winemakers resisted to adopt innovative ideas to treat grape disease (Stoupis, 2019).

Corfu is the second biggest Greek island in the Ionian Sea, between Greece and Italy, but by far the most popular tourist destination among Ionian islands (Hellenic Civil Aviation Authority, 2020; INSETE, 2020). In an effort to leave behind mass tourism and build alternative and niche tourism instead, the carrying capacity of wine tourism appears promising. There are two wine routes on the island, one in south-west Corfu and one in the north-west (Wines of Greece, 2023). According to recent data, wine production makes up for approximately 8% of agriculture on the island (Corfu Food and Wine Festival, 2018). Different grape varieties can be found on the south of the island and others on the north, whereas, at some villages, a limited production of different varieties is evident as well.

Out of the eight wineries which accepted to participate in the study, five were open to the public at the time of the study, whereas the other three were adequately equipped yet remained closed to visitors. Preliminary content analysis of the websites paired with responses to the structured questionnaire assisted in profiling the winemaker, the winery and assessing the winery's involvement in wine tourism activities.

Table 7.1 summarizes some key characteristics of the respondents. Table 7.2 describes the infrastructure and services available at those wineries that tourists may visit and indicates the languages in which the tour may take place. At the time of data collection, the Ministry of Tourism had not yet certified any of the wineries for wine tourism; in fact, only one winery was, at that point, in the process of applying for relevant certification.

In sum, half of the wineries follow trends in the wine tourism industry (e.g. by offering food and wine pairing), whereas most winery owners/managers have received formal education on tourism, agriculture, management or a combination of the above. Although involvement with wine tourism varies, it becomes obvious that only a limited number of wineries pursue a more holistic experience, ranging from the use of audiovisual equipment to offering a harvest experience or hosting and organizing events.

Perceptions of Sustainable Wine Tourism: The Island of Corfu

In-depth interviews reveal details on wineries' understanding of sustainable wine tourism development, thereby providing us with enough information to proceed with applying the Gioia *et al.* (2013) methodology for exploring the theory of dynamic capabilities in the case of Corfu. Preliminary analysis indicates that all respondents were familiar with the concept of sustainability, although none had received a formal sustainability certification or education. Quite often the wineries follow the principles of organic farming and generally act driven from their respect for the environment and product quality standards. In the words of C5: '[through our sustainable practices] what comes from the grape is returned to the grape' and C4: 'all farm activities complement each other. For instance, we use sheep manure, pruning from trees, residues from grapes and our annual olive oil production as fertilizer. Also, we use broad beans. (...) In January we sow broad beans, vetch, etc., and, towards the end of March, these will be milled and contribute to what we call green fertilization. This enriches the soil'. Wine producers adopt sustainable practices driven by their wish to protect the vineyard, control production and the end product. Respondents who are not involved with sustainable practices reveal that they avoid such practices because of time limitations, high cost and limited know-how.

The Gioia methodology allowed us to identify the theoretical dimensions which relate to the opportunities and challenges of wine tourism, and feed into the theory of dynamic capabilities (Gioia *et al.*, 2013). Four conceptual dimensions emerged in terms of opportunities:

- The culture of Corfu, local traditions and products can be the cornerstone for wine tourism development. This implies highlighting the significance of local cuisine and pairing traditional dishes with unique local wines. This is how knowledge and interest will accumulate (Alonso *et al.*, 2020).
- Establishing wine tourism requires: (i) intensive production of local grape varieties; (ii) ensuring these are not threatened with extinction; and (iii) empowering the wine production of Corfu.
- Progress in extant sustainable wine practices and its future prospects. Wine producers prefer sustainable practices and are interested in sustainability both in principle and in terms of its benefit on their end products.
- Succession links to the interest of subsequent regenerations, new innovative ideas, use of accumulated family knowledge and experience, and professional agreements helping to position Corfu as a wine destination.

Similarly, three conceptual dimensions emerged from challenges:

- The danger of extinction of local grape varieties, which all winemakers acknowledge: without grapes, it is impossible to produce sufficient amounts of local products. Thus, sustainable wine tourism development cannot be achieved.
- Synergies among winemakers and between winemakers and local authorities are deemed necessary in order to achieve wine tourism development. Isolated efforts should be substituted with a common vision, whereas local administration can

Table 7.1. Profile of winery owners/managers.

Code	Position	Wine business experience (in years)	Gender	Year of birth	Highest degree	No. of children	Educated in			Following trends in the wine industry
							Agriculture	Tourism	Management	
C1	Owner	30	Male	1983	University	0	–	–	–	No
C2	Owner	14	Male	1992	University	0	X	X	X	Yes
C3	Owner	40	Male	1966	University	2	X	–	–	No
C4	Manager	10	Male	1976	University	1	–	X	X	Yes
C5	Owner	10	Female	1975	High school	2	X	X	X	Yes
C6	Owner	30	Male	1977	University	2	–	–	–	Yes
C7	Owner	32	Male	1956	College	2	X	–	–	No
C8	Owner	30	Male	1955	College	2	–	–	–	No

Table 7.2. Profile of wineries equipped for wine tourism (infrastructure and services available).

Code	Infrastructure available							Languages in which a tour can be offered
	Signage	Parking	Room for wine tasting	Events hall	Store	Restaurant	Museum	
C3	–	X	X	–	X	–	–	EN
C4	X	X	X	X	X	–	X	EN, IT, DE
C5	X	X	X	X	X	–	X	EN, IT
C6	X	X	–	–	X	–	–	EN, SPA, FR
C7	X	X	X	–	–	–	–	EN, FR, RU

Code	Services available							
	Winery tour	Vineyard tour	Wine tasting	Food and wine pairing	Audiovisual	Harvest experience	Hosting events	Organizing events
C3	X	X	X	X	–	X	–	–
C4	X	–	X	X	X	–	X	X
C5	X	X	X	X	–	X	X	X
C6	X	X	X	–	X	–	–	–
C7	X	X	X	X	–	–	–	–

Table 7.3. Gioia methodology feeding into the theory of dynamic capabilities (adapted for the case of Corfu from Alonso *et al.* (2020).

Code	Sensing	Seizing	Reconfiguring
C1	X	X	–
C2	X	X	X
C3	X	–	–
C4	X	X	X
C5	X	X	X
C6	X	X	X
C7	X	–	–
C8	X	–	–

support in terms of communication, promotion and funding.

- Cost emerges as a major challenge. The cost to pursue sustainability certifications is high compared with the net profit of these family businesses. Equally high is the cost to develop organized wine tourism activities.

Table 7.3 summarizes how sustainable wine tourism development on Corfu reflects on Teece *et al.*'s (1997) theory of dynamic capabilities. Table 7.3 particularly reveals that sensing is evident for all winemakers; that is, all wineries recognize the benefits of sustainable wine tourism, as well as the opportunities and challenges, although they may not equally prioritize sustainability principles. Focusing on seizing the opportunities, C1 acknowledges the significance of local varieties but hesitates to invest in wine tourism activities, pointing out that 'on Corfu, there aren't many grapes. More needs to be planted. We now start planting more and we'll see how it goes. We are just trying to expand the grape production at the moment and see how it goes. (...) It takes a lot of money to meet all the standards and open the winery to visitation'. Reconfiguring the opportunities is not that strong on Corfu. In fact, only four winemakers follow developments in wine management and consumers' changing needs. Some winemakers are happy with solely producing and selling wine, whereas others attribute their non-involvement with wine tourism to time constraints; in the words of C7: 'wine tourism needs in-depth interest/[sense of] dedication, enthusiasm, gusto [i.e. sense of taste], passion [*meraki* in Greek], available time and education, and all three are largely absent on the island'.

Applying the Theory of Dynamic Capabilities to Explore Sustainable Wine Tourism Development

This chapter set off to investigate the application of the theory of dynamic capabilities in the pursuit of sustainable wine tourism development in the case of Corfu. The aforementioned results helped us to identify that, when focusing on dynamic capabilities, what emerges as opportunities include the need to further progress, establish and continue the efforts towards sustainable tourism development and support the culture of Corfu as an emblem. Such opportunities suggest that Corfu is currently at an emerging stage in its wine tourism life cycle (Deery *et al.*, 2012; Alebaki and Koutsouris, 2019). The conceptual dimensions included in Fig. 7.1 summarize those opportunities that await wineries to harvest. Similarly, the dashed lines on Fig. 7.1 highlight those challenges which do not facilitate Corfu reach its full potential as a sustainable wine destination. At the same time, the limited reconfiguration of opportunities suggests stagnation in terms of harvesting future prospects. Still, this is not necessarily an obstacle that cannot be tackled.

In detail, Fig. 7.1 summarizes the facts that: (i) a holistic involvement of the wineries with wine tourism is more important than the exact number of wineries open to the public; (ii)

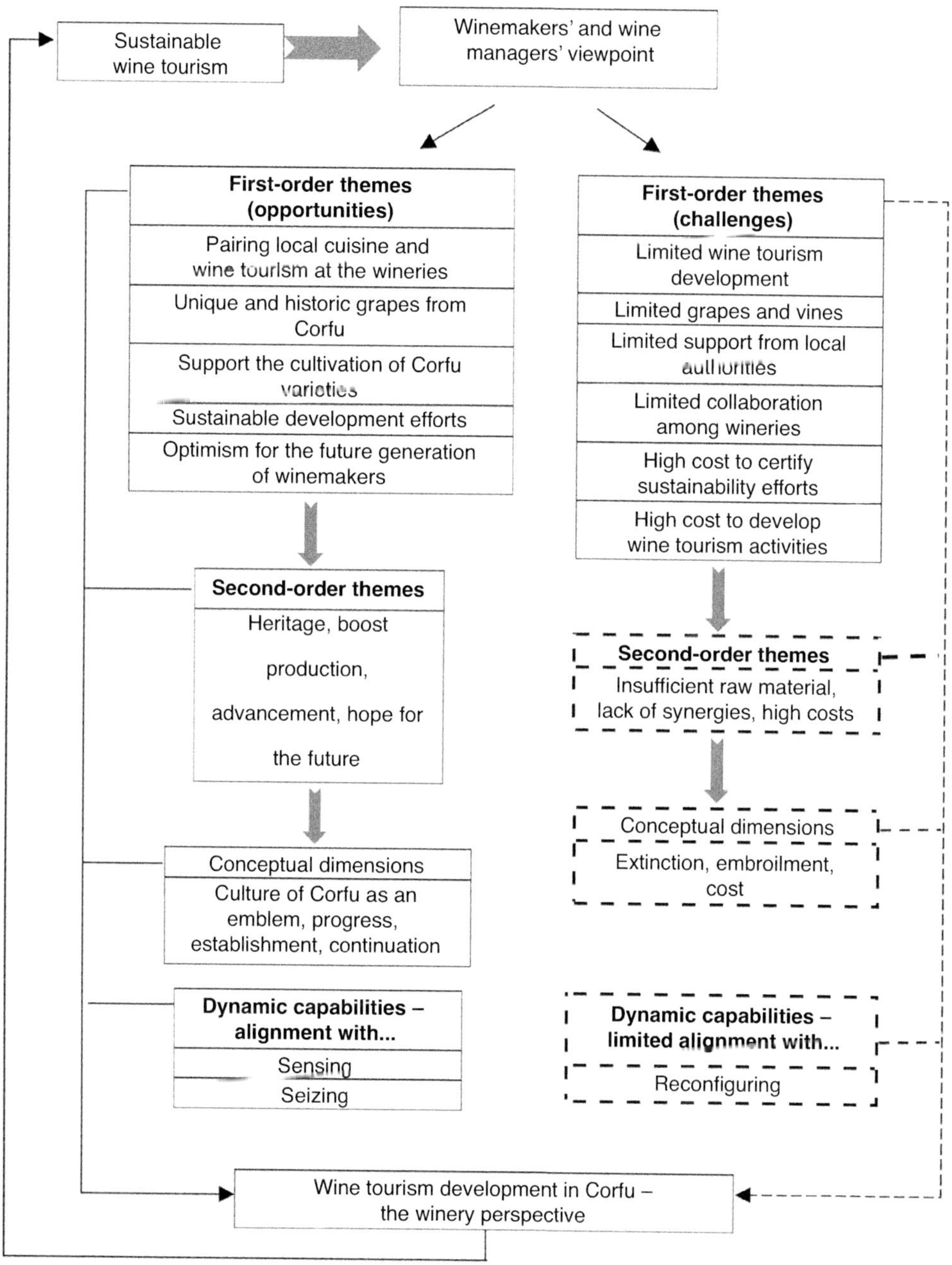

Fig. 7.1. Conceptual framework summarizing the findings of the Gioia *et al.* (2013) methodology and the theory of dynamic capabilities (Teece *et al.*, 1997; Alonso *et al.*, 2020) in the case of sustainable wine tourism development.

winemakers' knowledge, training and briefing are vital to build up a competitive advantage; (iii) adequate planning and collaborations are necessary to pursue sustainable wine tourism development; and (iv) local products of high quality are critical for the wine tourism experience. Thus, winemakers need to decide whether they wish their winery to be part of the wine tourism experience, or their superficial involvement may negatively impact other wineries and the wine tourism image of the region (Sparks and Malady, 2006). Moreover, knowledge is key to establish a competitive advantage in family businesses (Cabrera-Suárez *et al.*, 2001) and, when passed on to subsequent generations, may accumulate more benefit. Besides, this learning process helps wineries learn how to innovate, listen to the market and adjust their business accordingly (Bjuggren and Sund, 2001). Still, when small family wineries make up for the majority in an area, the collaboration with each other and other public and private stakeholders in the area is necessary in order to develop wine tourism in the area (Alonso and Liu, 2010). This collaboration should come as the result of adequate, strategic planning and should consider the fact that local grape varieties can attract visitors (Thiene *et al.*, 2013; Woods *et al.*, 2015; Soulek *et al.*, 2016). Besides, local wines and the winescape, which embraces local heritage and culture, constitute significant travel motives and are important in the pursuit of a competitive advantage (Howley *et al.*, 2012).

Conclusion, Implications and Future Directions

The current chapter focused on a well-established mass destination which now strives to develop niche tourism and particularly address wine tourists through sustainable wine tourism development. In specific, we hereby collected and analysed interview data from those wineries located on the island of Corfu in Greece that were equipped to accept wine tourists, even if some of them preferred keeping their doors closed and focusing on the production and exports rather than engaging in wine tourism activities. Content analysis of winery websites, a structured questionnaire and additional insights of wine specialists on Corfu helped us employ the Gioia *et al.* (2013) methodology to analyse empirical results and apply Teece *et al.*'s (1997) model of dynamic capabilities in the case of pursuing sustainable wine tourism. Analysis reveals the importance of sensing, seizing and configuring opportunities and challenges in order to achieve sustainable wine tourism development.

Earlier studies have investigated dynamic capabilities in the wine industry in the case of New-World wine countries (e.g. Peru) and Old-World wine countries (i.e. Italy and Spain) which are more developed as wine destinations (Alonso *et al.*, 2019, 2020). Thus, theoretical implications of this chapter connect with providing insights of an island destination that is a well-established leisure destination but only at the developing stage of its life cycle as a wine destination. Opportunities, challenges and subsequent similarities, differences, conceptual dimensions and dynamic capabilities relate to such characteristics (i.e. a developing island wine destination) which remain largely unexplored. Such findings confirm the importance of small family wineries for the industry and regional sustainability (Vukovic *et al.*, 2020), yet further suggest the need for synergies to extend beyond the wine sector and into other food and cultural aspects of the island. In extent, in an effort to see how the theory of dynamic capabilities develops, future studies could incorporate the views of other stakeholders, less directly connected to the wine industry per se (e.g. local authorities, restaurants and other business owners, residents).

In order to shed light on possible practical implications, it is important to acknowledge that most winemakers recognize both opportunities and challenges for sustainable wine tourism development on Corfu. This fact, as reflected also in how winemakers sense extant opportunities and appear motivated rather than discouraged from the challenges, suggests a promising future for wine tourism on Corfu and its sustainable development. The theoretical dimensions reflecting opportunities particularly indicate how sustainable wine tourism development can take place only with respect to local wine heritage and culture. On the one hand, the challenges faced reveal those characteristics on which both the wineries and the wine sector in general need to work (Alonso *et al.*, 2020); that is, with the local grape varieties going extinct and the economic and financial discrepancy of the wineries, no wine tourism is feasible. The theoretical dimensions reflecting opportunities indicate how sustainable wine tourism development can take place only with respect to local wine heritage and culture. The theoretical framework of Fig. 7.1 offers clear practical implications and helps stakeholders in the case of developing wine destinations to understand and harvest the benefits of sustainable development of wine tourism.

Sensing and seizing opportunities and blending such dynamic capabilities with the opportunities recognized is crucial for sustainable wine tourism development. Still, in established leisure yet developing wine destinations attention needs to be paid due to limited reconfiguration. Future research needs to explore reconfiguration in more detail in terms of its meaning, particularly for emerging and island wine destinations. Limited reconfiguration seems to lead wineries to repetition and those wine tourism strategies that are established and well-known wineries and wine destinations follow, instead of facilitating the flow of adequate information and the development, innovating thinking and subsequent creation of a unique selling proposition relevant to the focal destination and winery. In the end, the current status of dynamic capabilities suggests the importance of wine tourism for wineries and the tourism sector alike and highlights the importance of extended synergies even in the case of such a niche tourism type as wine tourism.

References

Alebaki, M. (2012) Factors affecting the development of wine tourism. PhD thesis, Aristoteleio University of Thessaloniki, Thessaloniki, Greece.

Alebaki, M. and Iakovidou, O. (2010) Initiatives towards wine tourism development in Greece. *Enometrica* 3(2), 15–24.

Alebaki, M. and Koutsouris, A. (2019) Wine tourism destinations across the life-cycle: a comparison of northern Greece, Peloponnese and Crete. In: Sigala, M. and Robinson, R.N.S. (eds) *Wine Tourism Destination Management and Marketing*. Palgrave Macmillan, Cham, Switzerland, pp. 463–482.

Alonso, A.D. and Liu, Y. (2010) Wine tourism development in emerging Western Australian regions. *International Journal of Contemporary Hospitality Management* 22(2), 245–262.

Alonso, A.D., Kok, S.K. and O'Brien, S. (2019) Understanding approaches to innovation through the dynamic capabilities lens: A multi-country study of the wine industry. *International Journal of Innovation Management* 23(6), 1950054.

Alonso, A.D., Kok, S. and O'Brien, S. (2020) Sustainable wine tourism development through the lens of dynamic capabilities and entrepreneurial action: an exploratory four-region perspective. *Tourism Recreation Research* 45(3), 401–419.

Barreto, I. (2010) Dynamic capabilities: a review of past research and an agenda for the future. *Journal of Management* 36(1), 256–280.

Bjuggren, P.O. and Sund, L.G. (2001) Strategic decision making in intergenerational successions of small- and medium-size family-owned businesses. *Family Business Review* 14(1), 11–23.

Cabrera-Suárez, K., De Saá-Pérez, P. and García-Almeida, D. (2001) The succession process from a resource- and knowledge-based view of the family firm. *Family Business Review* 14(1), 37–46.

Cepeda, G. and Vera, D. (2007) Dynamic capabilities and operational capabilities: a knowledge management perspective. *Journal of Business Research* 60(5), 426–437.

Corfu Food and Wine Festival (2018) *The Wine of Corfu*. Available at: https://www.corfufestival.org/2018/03/07/the-wine-of-corfu/ (accessed 6 December 2023).

Deery, M., Barry O'Mahony, G. and Moors, R. (2012) Employing a lifecycle typology to generate a unified and strategic approach to regional wine tourism development. *Tourism Planning & Development* 9(3), 291–307.

Eisenhardt, K. and Martin, J. (2000) Dynamic capabilities: what are they. *Strategic Management Journal* 21(10–11), 1105–1121.

Eriksson, T. (2014) Processes, antecedents and outcomes of dynamic capabilities. *Scandinavian Journal of Management* 30(1), 65–82.

Flores, S.S. (2018) What is sustainability in the wine world? A cross-country analysis of wine sustainability frameworks. *Journal of Cleaner Production* 172, 2301–2312.

Foley, G. and Timonen, V. (2015) Using grounded theory method to capture and analyze health care experiences. *Health Services Research* 50(4), 1195–1210.

Forbes, S.L., Cohen, D.A., Cullen, R., Wratten, S.D. and Fountain, J. (2009) Consumer attitudes regarding environmentally sustainable wine: an exploratory study of the New Zealand marketplace. *Journal of Cleaner Production* 17(13), 1195–1199.

Getz, D. and Brown, G. (2006) Critical success factors for wine tourism regions: a demand analysis. *Tourism Management* 27(1), 146–158.

Gioia, D.A., Corley, K.G. and Hamilton, A.L. (2013) Seeking qualitative rigor in inductive research: notes on the Gioia methodology. *Organizational Research Methods* 16(1), 15–31.

Getz, D., Dowling, R., Carlsen, J. and Anderson, D. (1999) Critical success factors for wine tourism. *International Journal of Wine Marketing* 11(3), 20–43.

Hall, C.M. and Mitchell, R. (2000) Wine tourism in the Mediterranean: a tool for restructuring and development. *Thunderbird International Business Review* 42(4), 445–465.

Greek Gastronomy Guide (2021) Gastronomy and Wine Tourism. Available at: http://www.greekgastronomyguide.gr/kerkyra/ (accessed 22 March 2022).

Hall, C.M. and Weiler, B. (1992) Introduction. What's special about special interest tourism?. In: Weiler, B. and Hall, C.M. (eds) *Special Interest Tourism*. Bellhaven Press, London, pp. 1–14.

Helfat, C.E. and Peteraf, M.A. (2003) The dynamic resource-based view: capability lifecycles. *Strategic Management Journal* 24(10), 997–1010.

Helfat, C.E. and Winter, S.G. (2011) Untangling dynamic and operational capabilities: strategy for the (n) ever-changing world. *Strategic Management Journal* 32(11), 1243–1250.

Hellenic Civil Aviation Authority (2020) Statistical data of Kerkyra Airport 'I. Kapodistrias' for the period 1994–2018. Available at: http://www.ypa.gr/en/our-airports/kratikos-aerolimenas-kerkyras-i-kapodistrias-kakk (accessed 22 January 2022).

Howley, P., Donoghue, C.O. and Hynes, S. (2012) Exploring public preferences for traditional farming landscapes. *Landscape and Urban Planning* 104(1), 66–74.

INSETE (2020) Report on Competitiveness and Structural Adaptation of the Wine Sector for the Year 2019 (in Greek). Available at: https://insete.gr/wp-content/uploads/2020/12/21-01_Ionian_Islands.pdf (accessed 22 November 2021).

Kim, H. and Bonn, M.A. (2016) Authenticity: do tourist perceptions of winery experiences affect behavioral intentions. *International Journal of Contemporary Hospitality Management* 28(4), 839–859.

Lane, B. (1994) What is rural tourism? *Journal of Sustainable Tourism* 2(1–2), 7–21.

Mastroberardino, P., Calabrese, G., Cortese, F. and Petracca, M. (2020) Sustainability in the wine sector: an empirical analysis of the level of awareness and perception among the Italian consumers. *British Food Journal* 122(8), 2497–251.

McMullen, J.S. and Shepherd, D.A. (2006) Entrepreneurial action and the role of uncertainty in the theory of the entrepreneur. *Academy of Management Review* 31(1), 132–152.

Montella, M. (2017) Wine tourism and sustainability: a review. *Sustainability* 9(1), 113.

Nave, A., Laurett, R. and do Paço, A. (2021) Relation between antecedents, barriers and consequences of sustainable practices in the wine tourism sector. *Journal of Destination Marketing & Management* 20, 100584.

Pandza, K. and Thorpe, R. (2009) Creative search and strategic sense-making: missing dimensions in the concept of dynamic capabilities. *British Journal of Management* 20, S118–S131.

Piening, E.P. and Salge, T.O. (2015) Understanding the antecedents, contingencies, and performance implications of process innovation: a dynamic capabilities perspective. *Journal of Product Innovation Management* 32(1), 80–97.

Poitras, L. and Getz, D. (2006) Sustainable wine tourism: the host community perspective. *Journal of Sustainable Tourism* 14(5), 425–448.

Serra-Cantallops, A., Ramón-Cardona, J. and Vachiano, M. (2021) Increasing sustainability through wine tourism in mass tourism destinations: The case of the Balearic Islands. *Sustainability* 13(5), 2481.

Soulek, K., Dodd, T. and Velikova, N. (2016) *Consumer Attitudes toward Texas Wine: A Comparative Report 2006–2016*. Texas Wine Marketing Research Institute, College of Human Sciences, Texas Tech University, Lubbock, Texas.

Sparks, B. and Malady, J. (2006) Emerging wine tourism regions: lessons for development. In: Carlsen, J. and Charters, S. (eds) *Global Wine Tourism: Research, Management and Marketing*. CAB International, Wallingford, UK, pp. 67–79.

Stasi, A., Muscio, A., Nardone, G. and Seccia, A. (2016) New technologies and sustainability in the Italian wine industry. *Agriculture and Agricultural Science Procedia* 8, 290–297.

Stoupis, N. (2019) The Corfu vineyard from Homer to this day (in Greek). Available at: https://www.corfuland.gr/el/istorika-kerkyra/corfu-history/o-kerkyraikos-ampelonas-apo-tin-epoxi-toy-omiroy-kai-toy-xenofonta.html (accessed 6 December 2021).

Teece, D.J. (2007) Explicating dynamic capabilities: the nature and microfoundations of (sustainable) enterprise performance. *Strategic Management Journal* 28(13), 1319–1350.

Teece, D.J. (2012) Dynamic capabilities: routines versus entrepreneurial action. *Journal of Management Studies* 49(8), 1395–1401.

Teece, D.J. (2014) The foundations of enterprise performance: dynamic and ordinary capabilities in an (economic) theory of firms. *The Academy of Management Perspectives* 28(4), 328–352.

Teece, D.J., Pisano, G. and Shuen, A. (1997) Dynamic capabilities and strategic management. *Strategic Management Journal* 18(7), 509–533.

Thach, L., Lease, T. and Barton, M. (2016) Exploring the impact of social media practices on wine sales in US wineries. *Journal of Direct, Data and Digital Marketing Practice* 17(4), 272–283.

Thiene, M., Galletto, L., Scarpa, R. and Boatto, V. (2013) Determinants of WTP for prosecco wine. *British Food Journal* 115(2), 279–299.

Vukovic, D.B., Maiti, M., Vujko, A. and Shams, R. (2020) Residents' perceptions of wine tourism on the rural destinations development. *British Food Journal* 122(8), 2739–2753.

Williams, P.W. and Dossa, K.B. (2003) Non-resident wine tourist markets: implications for British Columbia's emerging wine tourism industry. *Journal of Travel & Tourism Marketing* 14(2–3), 1–34.

Wines of Greece (2023) PGI of Corfu (in Greek). Available at: https://winesofgreece.org/el/pgi/πγε-κέρκυρα/ (accessed 6 December 2023).

Woods, T.A., Deng, X., Nogueira, L. and Yang, S.H. (2015) Local wine expenditure determinants in the Northern Appalachian states. *Journal of Food Distribution Research* 46(2), 30–50.

8 A Comprehensive Outlook of Culture, Creativity and Sustainability in Rural Areas Through Tourism Development

Sıla Karacaoğlu*
Department of Tourism Guidance, Faculty of Applied Sciences, Bilecik Şeyh Edebali University, Bilecik, Türkiye

Abstract
One of the main components of the creative movement in tourism has been the experience economy. In particular, the market for unforgettable travel experiences is expanding as a rising number of travellers – known as creative tourists – seek out first-hand experiences as active participants in the creation process, as opposed to being passive recipients of experiences that have already been prearranged. In this context, creative tourism, which emerged with the blending of culture and creativity, is a new paradigm and niche tourism type through which 21st-century tourists try to find transformative experiences. Since creative tourism does not require large investments and utilizes existing cultural and natural elements, it is an important approach to overcoming poverty and economic problems and developing tourism in rural areas. However, it can be said that most of the research on creative tourism focuses on the urban context and there is not much work focused on how creativity can be used and developed in the rural tourism environment. From this point of view, this chapter aims to examine the role of this niche tourism type in rural sustainable development by addressing the concept of creative tourism in a comprehensive manner and to contribute to the literature by analysing creative tourism practices in rural areas.

Keywords: Culture, creative tourism, sustainable tourism and development, rural areas, Türkiye

Introduction

The problems caused by excessive tourism in numerous major cities worldwide challenge the notion that larger is superior. Additionally, tourists' aspirations for genuine, intimate experiences that involve local life, culture and knowledge are also at odds with this idea (Scherf, 2021). The need for places to differentiate themselves in an increasingly globalized world has heightened awareness of the difficulties confronting smaller communities. How can regions with sparse inhabitants or limited resources effectively compete against 'world cities' and other larger economic hubs that dominate the global economy? Small cities and rural communities have the capacity to utilize information, relationships and creative resources to better position themselves in global networks and promote growth (Richards, 2019). In this context, creative tourism – derived from cultural tourism and considered its continuation (Palenčíková and Csapó, 2021) – is a prominent trend in the

*Corresponding author: sila.karacaoglu@bilecik.edu.tr

 Niche Tourism and Sustainability: Perspectives, Practices and Prospects (eds A. Farmaki *et al.*)
DOI: 10.1079/9781800626669.0008

global tourism market. It involves the utilization of creative knowledge and skills through shared and co-created activities which are inherent to the distinctive characteristics of the destination (Baixinho *et al.*, 2020).

Creative tourism is a novel and specialized form of tourism that has evolved from the combination of culture and creativity. It involves 21st century tourists seeking transformative experiences. The focus of this concept is on the artistic and cultural activities that are exclusive to a particular holiday destination. It allows guests to actively participate in developing their skills and expressing their creativity (Castro *et al.*, 2020). The market for remarkable travel experiences is growing as an increasing number of travellers, referred to as creative tourists, actively seek out first-hand experiences and participate in the creation process rather than passively receiving prearranged experiences (Bakas *et al.*, 2019). In recent years, there has been a significant increase in the number of courses offered in fields such as languages, gastronomy, art and photography. This growth is driven by both the strong demand for creative talents and the increasing number of creative producers who are catering to this demand (Richards, 2011). In addition, creative tourism is closely linked to the creative economy and offers substantial benefits at both economic and social levels. These advantages include the development of innovative tourism products, experiences and destinations, enhanced competitiveness and reputation of the destination, access to new customer segments, promotion of sustainable tourism and the growth of the local community (Palenčíková and Csapó, 2021; Sava, 2021).

Considering all these effects, it can be acknowledged that creative tourism is an important niche tourism form that enhances and sustains the tourist appeal of rural destinations. This chapter seeks to investigate the impact of this specific form of tourism on the sustainable development of rural areas. It does so by thoroughly exploring the concept of creative tourism and analysing its implementation in rural regions in Türkiye. Specifically, the chapter aims to contribute to existing theory and practice by drawing from three cases in rural areas.

Theoretical Background

Emergence, development and definition of creative tourism

The concept of creative tourism was first introduced as a type of tourism by Pearce and Butler (1993); however, no specific definition was provided (Richards, 2011). The concept has become clearer and more developed through the studies conducted in several regions of Europe in the 1990s. Some partners within the Eurotex Project, which aimed to revitalize craft products through tourism in order to attract tourists to the local culture, started to provide more opportunities to learn about local products. This project facilitated the development of a more meaningful and impactful tourism product by enabling visitors to establish a more holistic connection with the communities they visit, particularly with local producers (Zoğal and Emekli, 2017). Following, the term *creative tourism* was introduced by Richards and Raymond (2000, p. 18), defined as 'tourism which offers visitors the opportunity to develop their creative potential through active participation in courses and learning experiences which are characteristic of the holiday destination where they are undertaken'.

The Creative Cities Network of the United Nations Educational, Scientific and Cultural Organization (UNESCO), which began in 2004, is focused on creativity. This programme emphasizes that creative tourism should not focus on the tangible cultural heritage, but also recognize the significant value of the intangible cultural characteristics of the destination (Ajanović and Çizel, 2015). The concept of creativity was emphasized during the conference titled 'Towards Sustainable Strategies for Creative Tourism', which was attended by tourism professionals, intellectuals and representatives of the UNESCO Creative Cities Network in the USA in 2006 (Zoğal and Emekli, 2017). According to UNESCO (2006) definition of the Creative Cities Network, 'Creative tourism is travel directed toward an engaged and authentic experience, with participative learning in the arts, heritage, or special character of a place, and it provides a connection with those who reside in this place and create this living culture'. In 2011, the concept was further developed with the first official creative tourism project initiated by Raymond in New Zealand. Raymond summarized the concept of creative tourism with the idea of Confucius: 'I hear and I forget, I see and I remember, I do and I understand' (Zoğal and Emekli, 2017). Indeed, in today's world, tourists have a greater inclination towards engaging in tourism activities, alongside the traditional practice of visiting aesthetically pleasing locations (Sungsuwan, 2018).

The emergence of a form of tourism called 'creative tourism' is the result of tourists becoming less satisfied with traditional forms of tourism, where local and traditional spaces are passively visited and symbolic structures are passively consumed, placing more importance on creativity and self-expression (Akdu, 2019). Creative tourism emerges from the combination of various products and services, resulting in novel offerings such as crafts tourism, art festivals, workshops and master classes. In these experiences, tourists are actively engaged and given the opportunity to become participants, allowing them to create lasting memories, acquire knowledge and develop skills (Richards and Wilson, 2007). Creative tourism encompasses a variety of place-based activities that involve artistic pursuits such as photography, painting, drawing, winemaking, cookery, stone or wood carving, ceramics, sculpting, flower arranging, crafts, dance, music and drama. Furthermore, it also encompasses the act of investigating and articulating one's artistic capabilities on a vacation. The major emphasis is on the actions and their connection to oneself, with the context or setting being of secondary importance (Smith, 2016). In other words, creative tourism offers tourists activities that include a process rather than a final product (Gordin and Matetskaya, 2012).

The relationship among creativity, culture and tourism

In recent years, the emergence of global trends has increased interest in new and different tourist tendencies, counteracting uniformity. After the 2000s, cultural tourism practices that emphasize 'living like a local' as a result of the experience and sharing economy have gained attention, highlighting active participation and creativity rather than passive visiting (Zoğal and Emekli, 2017). In this context, researchers have focused on studies centred around 'creative tourism'. The topic of creative tourism research is a dynamic and emerging area that draws knowledge from various disciplines such as tourism, cultural heritage, cultural development, creativity, cultural geography, and local and regional development (Bakas *et al.*, 2021).

Undoubtedly, culture, creativity and tourism share a symbiotic relationship as they reciprocally amplify and bolster each other. Culture plays a vital role in the visitor experience by adding uniqueness to a place, making it more attractive and giving it a distinct character (Sattar, 2018). Experiential value is derived from innovative cultural tourism and rural tourism, which allow passengers to fully engage in a location that offers unique activities and symbolic significance (Alvarez, 2010). The correlation between creativity and culture has led to the emergence of a novel framework in the tourism sector, impacting both supply and demand and revealing a distinct visitor profile. The motivation to travel is a crucial factor that distinguishes cultural tourism from other forms of tourism. Cultural tourism is driven by various reasons, one of which is the desire to explore and gain a deeper understanding of the local culture through its cultural heritage. This not only benefits the tourism industry but also contributes to the preservation and promotion of the cultural heritage itself. Tourists are drawn to the cultural heritage, and, in turn, tourism has a role in safeguarding the legacy so that it can be utilized by travellers (Sava, 2021).

Historically, tourism was often regarded as a detrimental influence on local crafts and traditions due to the perceived changes brought about by the process of touristification. Nevertheless, there is a shift in perspective regarding tourism, with a growing emphasis on highlighting local history, especially intangible heritage. This shift aims to use tourism as a catalyst for reviving cultural practices and art forms (Bakas *et al.*, 2019). The need for innovative tourism experiences has been increasing within the framework of the 'experience economy' or 'educational tourism'. Presently, travellers are seeking more participatory and rewarding experiences rather than simply being catered to by the travel industry (Ferreira *et al.*, 2019). Creative tourism is an innovative approach to cultural tourism that encompasses four key elements: active engagement, expressive creativity, educational experiences and integration with the local people (Duxbury, 2020). Within the realm of creative tourism, the focus moves away from physical aspects of culture towards more abstract elements. The fundamental aspect of this type of tourism involves the sharing of information and abilities between the host and the guest. This results in a cultural tourism that is more focused on the local community, fairer and potentially more genuine in nature (Djukic and Vukmirovic, 2012). On the other hand, creative tourism offers distinct benefits compared with cultural tourism. First and foremost, as the capital of creative tourism is formed by the tourists themselves, the outcomes are unquestionably authentic in the perspective of the tourists. As tourists create their own experiences, they combine physical resources with abstract resources that ultimately stem from their views, previous experiences and aspirations. Subsequently, a process of innovative consumption and innovative manufacturing takes place. These intangible resources, referred to as creative and social capital, are utilized by individuals to achieve personalized outcomes. These outcomes are subsequently incorporated into the individual's existing creative and social capital (Richards and Wilson, 2006). In summary, creative tourism can be considered as an extension of cultural tourism that elevates it to higher levels (Richards, 2011).

Creative tourism as a sustainable tool for revitalizing rural destinations

Cultural tourism, which initially might be described as primarily traditional and limited to visits to prehistoric and ancient cultures, has transformed into a cultural tourism model that integrates traditional, contemporary and living cultures since the 1980s. After the 1990s and 2000s, the phenomenon of massification in the contemporary era began to be criticized, leading to a significant focus on alternative approaches and the paradigm of postmodern tourism within the framework of sustainability in tourism research. In other words, tourism has rapidly moved away from its mass nature in the modern era and has evolved towards individualism (Zoğal and Emekli, 2017). In contemporary times, creativity is regarded as a strategic factor for sustainable tourism management (Ajanovïc and Çizel, 2015). States, regions and cities aim to ensure the long-term viability of their economy by establishing sectors that allow them to maintain competitiveness. These sectors should align with the principles and objectives of sustainable development across all key areas, as well as with the advancement of information and communication technologies (Djukic and Vukmirovic, 2012).

As worldwide efforts to promote sustainable development models increase, adopting a de-growth paradigm is seen as advantageous in mitigating the downsides of tourism. This context emphasizes the need for creating pathways to develop smaller and more advantageous tourism models. These models should prioritize the unique characteristics of each location, cultural sustainability, social inclusion and respect for the sociocultural values of local communities. The aim is to improve the local quality of life through alternative tourism approaches. This concept has been discussed by Korez-Vide (2013) and Duxbury *et al.* (2021). Despite the importance of these variables, economic sustainability remains the primary consideration for promoters, stakeholders and local communities (Qiu *et al.*, 2019). An essential aspect of sustainability is the incorporation of innovative methods into the tourism industry, which serves as a reliable method to distinguish the local tourist offerings from several other rural tourism options worldwide (Souca, 2019).

Rural areas are commonly defined by their picturesque scenery, peacefulness, distinct culture and the distinctiveness of their physical and non-physical local resources. However, they are also recognized for the difficulties they face, such as a decreasing population and stagnant economy (Gato *et al.*, 2020). This scenario is when the concept of sustainable tourism becomes relevant. It functions as a remedy that will invigorate local economies and save cultural and ecological assets. Rural development aims to rejuvenate these areas by providing tourists with an authentic and immersive experience, such as engaging in ecotourism. The goal is to preserve the traditional way of life while also creating a tourism destination (Funduk *et al.*, 2024). In order to achieve sustainable cultural tourism, the management of both intangible and tangible cultural heritage with tourism activity must prioritize the involvement of stakeholders and local communities. This ensures that both the cultural heritage asset and the people benefit from the decision-making processes (Duxbury *et al.*, 2021). Creative tourism is a valuable strategy for addressing poverty and economic challenges, as well as promoting tourism in rural regions, due to its low investment requirements and utilization of existing cultural and natural resources (Einali *et al.*, 2019). Nevertheless, the majority of research on creative tourism mostly concentrates on the urban setting, with limited attention given to exploring the utilization and cultivation of creativity in rural tourism environments (Souca, 2019).

The emergence of creative tourism is apparent in both rural areas, where innovation is necessary to address the absence of economic options, and in cities, which are regarded as the driving forces behind the creative economy (Richards, 2011). The development of creative tourism in rural areas offers numerous benefits including the revitalization of local crafts and traditions, fostering a sense of cultural pride among the population, preserving both tangible and intangible aspects of culture, generating new employment opportunities and creative industries, stimulating economic growth through the spillover effects of creativity and enhancing the competitiveness of the destination (Djukic and Vukmirovic, 2012). In rural areas, artisan entrepreneurship serves as a means for the local community to acquire knowledge and skills from talented individuals. When handicraft production is combined with tourism, the entrepreneurial venture is more likely to generate profits (Bakas *et al.*, 2019; Souca, 2019; Teixeira and Ferreira, 2019). The objectives of tourism development include the creation of novel tourism offerings, experiences and destinations, enhancing the competitiveness and reputation of the destination, expanding the target market to attract new clients, promoting sustainable tourism practices and fostering the growth of the local community (Palenčíková and Csapó, 2021). In addition to this, creative resources possess greater sustainability, renewability and mobility compared with tangible resources. Additionally, creativity entails the active generation of value by both providers and consumers, resulting in their collaboration as co-producers (Carvalho *et al.*, 2016).

In summary, creative tourism is a contemporary and environmentally friendly form of cultural tourism that focuses on utilizing the cultural heritage of destinations to both safeguard and generate economic advantages for the local community. Additionally, it aims to provide tourists with authentic holiday experiences (Sava, 2021). However, there are also some debates in the literature regarding this matter. From a tourist's standpoint, actively participating in the cultural activities of a destination undoubtedly enhances one's overall experience. To what extent may creative tourism be developed? Empirical evidence demonstrates that numerous local customs have been altered to cater to the desires of tourists, resulting in the creation of artificial traditions or the fabrication of authenticity. This phenomenon is further compounded by contemporary advancements in tourism development. Therefore, should tourist involvement activities be abstained from or even prohibited if they alter or encroach upon the local culture (Jelinčić and Žuvela, 2012)? What types of creative tourist activities should be promoted to encourage sustainable tourism? At this point, the involvement of local stakeholders is crucial for developing suitable standards and strategies for managing creative tourism. Furthermore, local tourism stakeholders should have a venue where they may express their opinions. Thus, it may be easier for individuals to have a sense of ownership over creative tourism activities on a societal level (Sungsuwan, 2018).

When analysing the examples in the literature, it is fair to state that creative tourism contributes to sustainable tourism development (Solène, 2011; Korez-Vide, 2013; Chugh, 2018; Duxbury *et al.*, 2021; Sutarman *et al.*, 2023). Creative tourism is highly significant in protecting and perpetuating the natural, cultural and historical riches of a destination. Sustainable tourism can be seen as an important tool for revitalizing and enhancing the utilization of existing products, creating awareness about cultural and creative assets, generating economic value, developing creative techniques, creating employment and prioritizing the local community. Creative tourism serves sustainability by offering a combination of culture, entertainment and travel that allows tourists to experience different cultures, discover local cultural values and personally engage in certain cultural practices.

Cases of Creative Tourism Practices in Türkiye

Türkiye is a cultural mosaic characterized by its geographical location, historical depth and unique cultural structure that carries universal value. It is evident that the country, particularly the local folk culture, is a rich resource for the creative tourism sector. Furthermore, the concept of creativity in Türkiye has gained importance with the Creative Cities Network programme initiated by UNESCO in 2004. As UNESCO (2006), eight cities from Türkiye are part of UNESCO's Creative Cities Network. Gaziantep (joined in 2015), Hatay (joined in 2017) and Afyonkarahisar (joined in 2019) are included in the network of creative cities in the field of music, while Kütahya (joined in 2017) and Bursa (joined in 2021) are among the cities of craft and folk arts. Besides these, there are other destinations in Türkiye that are part of creative tourism activities. This section examines three of the practices of creative tourism in rural areas of Türkiye.

Pottery making: Avanos, Nevşehir

Cappadocia is a significant cultural tourism centre in Türkiye because of its historical, geographical and cultural background. Today's Cappadocia is mostly a tourist destination that encompasses the provinces of Nevşehir, Aksaray, Niğde, Kayseri and Kırşehir. In a narrow sense, Cappadocia represents the location situated inside the triangle of Nevşehir, Ürgüp and Avanos, which is significant in terms of tourism and the intensity of tourism activities. This situation is mainly due to the higher prevalence of fairy chimneys and geographical formations in this region, as well as the greater tourism activity seen in this area. The region of Cappadocia has hosted great empires and cultures such as the Hittites, Romans, Byzantines, Seljuks and Ottomans from the Neolithic period to the present day. The past is very significant, especially from the perspectives of Christianity and Islam (Erol, 2021).

When visiting Cappadocia, tourists tend to visit Avanos where the tradition of ceramics making has been going on since the Hittites, and today there are ceramic workshops on every corner. Tourists can participate in art and pottery workshops through organized tours or individually. In the workshops, visitors can interact with pottery masters, witness pottery-making demonstrations, delve into the history of pottery and its production process, and engage in their own pottery-making process by sitting at the pottery wheel, donning the traditional wide trousers worn by the masters. They can take their handmade pottery home as a souvenir. Tourists can not only have a pleasant and fun experience, but they can also learn about the traditional art of pottery. Tourists' interest in pottery-making experiences increases Avanos' touristic appeal and contributes to the preservation and transmission of Cappadocia's handicraft tradition to future generations.

Bread and yufka making: Kurşunlu village, Gölpazarı, Bilecik

The district of Gölpazarı, with a population of approximately 10,000, has developed in ecological agriculture and animal husbandry due to its clean water sources, soil richness, pastures and biological diversity. The village of Kurşunlu, located inside the district, possesses microclimatic characteristics due to its proximity to the Sakarya River coastline. Kurşunlu village has opened up to tourism under the leadership of women, with a focus on ecotourism. The tourism movement, initiated entirely by civilians and led by women, has primarily included approximately 30 travel agencies from Istanbul and surrounding provinces in tour programmes to Kurşunlu village and some nearby villages (Gülcan *et al.*, 2023).

Women in the village of Kurşunlu have transformed certain rooms in their homes into guesthouses for tourists. The village has the capacity to accommodate 35–40 tourists and provides breakfast, lunch and dinner to its guests. Additionally, they engage in the sharing of village life with tourists. Moreover, there are other activities that tourists can partake in. These activities are arranged for adult groups, as well as camps with a 'father–son, mother–daughter' theme in the hamlet. Visitors have the opportunity to gather vegetables including tomatoes, peppers, parsley and assorted fruits from the gardens of the accommodation they are staying at, in addition to collecting eggs from the chicken coop and milk from the livestock. Herbs that can be consumed are also gathered based on the time of year. Furthermore, guests have the opportunity to engage in the process of crafting bread and yufka, noodles, tomato paste, cheese, jam and molasses with the ladies residing in the hamlet. Compact farm vehicles are utilized for visiting neighbouring settlements on the plateau. During the vintage season, tourists are transported to adjacent villages. Annually, on 3 October, the community hosts a customary celebration where over a hundred flowers and heirloom seeds are exchanged. Children travellers are educated about ancestral seeds and then participate in planting them during the right seasons. During the evenings, tourists are informed about the traditions and customs of the village as well as their stories, that takes place in the village house. Additionally, a traditional night, known as henna night, is organized for the tourists. This event includes Balkan music and involves dressing the tourists in traditional clothing that belonged to the ancestors of the local people (Karacaoğlu and Yolal, 2021; Gülcan *et al.*, 2023).

As mentioned above, women in the village successfully integrate intangible cultural heritage and creative tourism practices. The most popular activity for tourists is to experience bread and phyllo (filo pastry) making with the women in the village. With the support of Türkiye, Azerbaijan, Kazakhstan, Kyrgyzstan and Iran, the making of thin bread (yufka and lavash) was added to UNESCO's Intangible Cultural Heritage List in 2016 (UNESCO, 2016). While yufka is made of unleavened dough and baked on a 'sac' placed on the fire to keep it intact for a long time, lavash is baked in stone or earth ovens made in traditional ways called 'tandır'. There are traditional practices in the preparation of thin bread that are based on sharing and collective labour. In these practices, which can be shown as an example of social solidarity, the bread is usually cooked for each family with the help of neighbours in a collective work method as a preparation for winter (Akdu and Boztoprak, 2023). Tourists can buy the products they make, especially bread and yufka, and take them as souvenirs. Thus, economic benefit is provided to the women in the village. In addition, by promoting and preserving their culture, women both leave a legacy to the next generations and ensure that tourists have unforgettable tourism experiences.

Çini (tile) making: Sırköy, Kütahya

The cities of İznik and Kütahya have been significant centres for the production of ceramic art, namely in the field of traditional Turkish arts, due to their ancient history. Kütahya is situated in the western region of Türkiye and has a population of 573,642 individuals. Çini, a revered creative heritage that has been passed down for ages, serves both cultural and social purposes within the community. Currently, there are numerous ceramic workshops in Kütahya that strive to continue the art of ceramics. There are around 435 craft workshops that employ a total of 15,000 craftsmen. Out of them, 95% are specifically involved in the manufacturing of çini, as reported by UNESCO in 2024. These workshops produce ceramics that are designed for everyday use and for tourism purposes such as plates, pitchers, bowls, vases, figurines and cups. In Kütahya, both traditional and Kütahya-style tiles are produced (Uzunget, 2019). Kütahya organizes numerous national and international events with the aim of promoting çini. The International Çini Contest, which attracted participants from 33 different nations in its most recent edition, is dedicated to the promotion and advancement of çini (UNESCO, 2024).

The Center for Creative Cultural Industries, currently being built in Kütahya, is set to be operational in the near future. The Sırköy project involves the construction of workshops in the centre district, where tourists can engage in crafts and traditional arts such as tile making, pottery, glitter wrapping and silver processing.

In addition, these workshops will manufacture and market products. The Sırköy workshops aim to contribute to the tourist and socio-economic development of the region by promoting the sustainability of tourism and cultural assets, as well as enhancing human capital (T.C. Kütahya Belediyesi, 2023).

Conclusions and Implications

Today, both tangible and intangible cultural and creative elements play a crucial role in enhancing the attractiveness of rural destinations. Successful creative tourism aims to offer travellers a valuable and distinctive experience that promotes local identities and values. Furthermore, it may be argued that creative tourism represents a novel form of tourism that focuses on appreciating and enhancing natural, cultural and human resources, rather than manipulating and exploiting them (Jelinčić and Žuvela, 2012). The effects of creative tourism include the opportunity to discover less popular areas, reduced expenses, avoidance of crowded destinations, the potential to sustain tourism in season-dependent locations, the generation of higher value through the activities offered and the employment of highly skilled workers (Sava, 2021). Creative tourism is continuously developing, as are the traits of new consumers and the deregulation of the tourism market. The impacts of overtourism, sustainability concerns and the emergence of cultural intermediators equipped with advanced communication tools, as well as the growing significance of local communities in the tourism industry, all contribute to this evolution. Further investigation is required to explore the potential of creative tourism as a viable alternative to the repetitive replication of culture, the absence of authenticity in tourism encounters, and the social, cultural and economic inequities in tourist destinations (Carvalho *et al.*, 2019).

This chapter examines the concept of creative tourism, which aims to promote economic and sociocultural sustainability in rural areas by integrating tourists with the local culture. It explores how tourists can live like locals, identify with the region and gain unique experiences. The chapter also discusses three examples of creative tourism activities in rural areas, specifically focusing on touristic attractions in Türkiye. Nevertheless, despite Türkiye's extensive cultural heritage spanning millennia, it might be argued that the country's creative tourism initiatives are inadequate. Given the fact that Türkiye attracts millions of tourists annually and holds a high position globally in terms of international tourist arrivals, it is worth considering the importance of promoting a niche form of tourism that can be enjoyed throughout the year, in addition to the popular 'sea–sun–sand (3S)' tourism product that Türkiye is largely known for. Creative tourism as an alternative form of tourism could emphasize the cultural and creative wealth of the country. It is evident in this situation that Türkiye's tourist strategy should make better use of creative tourism practices as a substitute for excessively developed forms of tourism. Furthermore, the influence of creative tourism in rejuvenating tourism in rural regions should be enhanced.

This chapter has considered the positive contribution of creative tourism as a niche type of tourism to sustainable tourism development in rural areas. However, some criticisms and concerns in the literature, such as practices that change local tradition and culture, loss of originality and authenticity, exploitation and commercialization of local culture, mass production of manual labour, uncontrolled tourism development and hostility of local people to tourists after a certain period of time, also need to be pointed out. Overall, when assessing the advantages of creative tourism through a sustainability lens, it can be concluded that creative tourism promotes the development of sustainable tourism. Creative tourism plays a crucial role in preserving and upholding the historical and cultural abundance of rural places. In addition to contributing to the existing literature, the chapter emphasizes the added value created in rural areas by transforming creative tourism practices in Türkiye, and similar destinations, into touristic products.

For creative tourism to be successful, destinations must design and produce experiences unique to their regions and avoid mass production. This can only be possible if the tourism stakeholders in the destination and/or the people and institutions involved in its management identify the areas where the destination has the potential for creativity and work on these areas in a planned manner. It can be ensured that all stakeholders, especially local people, are provided with the necessary training on creative tourism and then appropriate environments can be created in cooperation.

Creative tourism practices can be integrated into slow tourism and various types of sustainable tourism as a result of the work to be done in cooperation with local administrations, private businesses and the people of the region. Thus, the destination can provide motivation for tourists to visit. Creative tourism practices can help tourists to gain unforgettable and meaningful experiences by improving the quality of their visit experience and develop a positive image perception of the region. Similarly, it can also provide economic, social and cultural benefits for local people.

It is thought that academic studies on creative tourism are insufficient and the research gap in this field should be addressed. Moreover, future research should investigate ways to strengthen the contribution of creative tourism to the development of sustainable tourism in rural areas. It is thought that conducting these studies using phenomenological approaches such as the Delphi technique and interview method may be beneficial in revealing the opinions of different stakeholders in more depth and providing more practical suggestions for tourism planning and tourism policy.

References

Ajanovic, E. and Çizel, B. (2015) Unesco yaratıcı kentler ağı ile antalya kentinin değerlendirilmesi. *Akdeniz İnsani Bilimler Dergisi* 5(1), 1–16.

Akdu, S. (2019) Sürdürülebilir turizm gelişimi ve yaratıcı turizm. *Gümüşhane Üniversitesi Sosyal Bilimler Dergisi* 10, 132–139.

Akdu, U. and Boztoprak, B. (2023) Somut olmayan kültürel miraslara yönelik yaratici turizm uygulamalari: pasinler lavaşi örneği. *Erciyes Akademi* 37(4), 1535–1565.

Alvarez, M.D. (2010) Creative cities and cultural spaces: new perspectives for city tourism. *International Journal of Culture, Tourism and Hospitality Research* 4(3), 171–175.

Baixinho, A., Santos, C., Couto, G., Albergaria, I.S., Silva, L.S. *et al.* (2020) Creative tourism on islands: a review of the literature. *Sustainability* 12(24), 10313.

Bakas, F.E., Duxbury, N. and Vinagre de Castro, T. (2019) Creative tourism: catalysing artisan entrepreneur networks in rural Portugal. *International Journal of Entrepreneurial Behavior & Research* 25(4), 731–752.

Bakas, F.E., Duxbury, N. and Albino, S. (2021) Creative tourism microentrepreneurs in Portugal. In: Morais, D.B. (ed.) *Tourism Microentrepreneurship. Bridging Tourism Theory and Practice*, Vol. 12. Emerald Publishing Limited, Bingley, UK, pp. 79–91.

Carvalho, R., Ferreira, A.M. and Figueira, L.M. (2016) Cultural and creative tourism in Portugal. *PASOS Revista de Turismo y Patrimonio Cultural* 14(5), 1075–1082.

Carvalho, R.M.F., da Costa, C.M.M. and Ferreira, A.M.A.P. (2019) Review of the theoretical underpinnings in the creative tourism research field. *Tourism & Management Studies* 15(1SI), 11–22.

Castro, T.V.D., Duxbury, N., Silva, S., Bakas, F., Carvalho, C. *et al.* (2020) *Creative Tourism: Guide for Practitioners. How Do You Regenerate Communities and Places by Combining Culture, Tourism and Creativity*. Centre for Social Studies of the University of Coimbra, Coimbra, Portugal.

Chugh, S.R. (2018) Creative tourism: an endogenous approach to developing culturally and environmentally sustainable tourism. *International Journal of Hospitality and Tourism Systems* 11(1), 60–66.

Djukic, A. and Vukmirovic, M. (2012) Creative cultural tourism as a tool in regional development. *Technics Technologies Education Management* 7(4), 1768–1778.

Duxbury, N. (2020) Catalyzing creative tourism in small cities and rural areas in Portugal: The CREATOUR approach. In: Scherf, K. (ed.) *Creative Tourism in Smaller Communities: Place, Culture, and Local Representation*. University of Calgary Press, Calgary, Canada, pp. 27–60.

Duxbury, N., Bakas, F.E., Vinagre de Castro, T. and Silva, S. (2021) Creative tourism development models towards sustainable and regenerative tourism. *Sustainability* 13(1), 2.

Einali, J., Mohammadi Yeganeh, B. and Ghasemlou, H. (2019) The role of creative tourism in sustainable development of rural areas (Case study: Historic-cultural villages in North-west of Iran). *Journal of Research and Rural Planning* 8(2), 19–39.

Erol, G. (2021) Yaratıcı turizm: Cappadocia Bölgesine Yönelik durum değerlendirmesi ve öneriler. *Yönetim, Ekonomi ve Pazarlama Araştırmaları Dergisi* 5(2), 111–126.

Ferreira, J., Sousa, B.M. and Gonçalves, F. (2019) Encouraging the subsistence artisan entrepreneurship in handicraft and creative contexts. *Journal of Enterprising Communities* 13(1/2), 64–83.

Funduk, M., Biondić, I. and Simonić, A.L. (2024) Revitalizing rural tourism: a Croatian case study in sustainable practices. *Sustainability* 16(1), 31.

Gato, M.A., Costa, P., Cruz, A.R., Perestrelo, M. and Tomaz, E. (2020) Creative tourism as an alternative, CREATOUR as an example. In: *International Workshop: Rethinking Culture and Creativity in the Technological Era, Florence, Italy, 20–21 February 2020.*

Gordin, V. and Matetskaya, M. (2012) Creative tourism in Saint Petersburg: the state of the art. *Journal of Tourism Consumption and Practice* 4(2), 55–77.

Gülcan, B., Sakal, H.B., Karaçar, E. and Karacaoğlu, S. (2023) *Bilecik Turizm ve Tanıtım Yol Geliştirme Haritası*. Türkiye Turizm TanıTım ve Geliştirme Ajansı Yayınları, T.C. Kültür ve Turizm Bakanlığı, Türkiye.

Jelinčić, D.A. and Žuvela, A. (2012) Facing the challenge? Creative tourism in Croatia. *Journal of Tourism Consumption and Practice* 4(2), 78–90.

Karacaoğlu, S. and Yolal, M. (2021) The role of women's leadership in community-based tourism development: the case of Kursunlu Village, Türkiye. In: Sandeep, K.W. (ed.) *The Routledge Handbook of Community Based Tourism Management*. Routledge, London, pp. 119–129.

Korez-Vide, R. (2013) Promoting sustainability of tourism by creative tourism development: how far is Slovenia. *Innovative Issues and Approaches in Social Sciences* 6(1), 77–102.

Palenčíková, Z. and Csapó, J. (2021) Creative Tourism as a New Tourism Product in Slovakia. The Theoretical and Practical Analysis of Creative Tourism: Formation, Importance, Trends. Faculty of Central European Studies, Constantine the Philosopher University in Nitra, Slovakia, *Europica Varietas* 146, 13.

Pearce, D.G. and Butler, R.W. (1993) *Tourism Research: Critiques and Challenges*. Routledge, London.

Richards, G. (2011) Creativity and tourism: the state of the art. *Annals of Tourism Research* 38(4), 1225–1253.

Richards, G. (2019) Creative tourism: opportunities for smaller places. *Tourism & Management Studies* 15(SI), 7–10.

Qiu, H., Fan, D.X.F., Lyu, J., Lin, P.M.C. and Jenkins, C.L. (2019) Analyzing the economic sustainability of tourism development: evidence from Hong Kong. *Journal of Hospitality & Tourism Research* 43(2), 226–248.

Richards, G. and Raymond, C. (2000) Creative tourism. *ATLAS News* 23, 16–20.

Richards, G. and Wilson, J. (2006) Developing creativity in tourist experiences: a solution to the serial reproduction of culture. *Tourism Management* 27(6), 1408–1413.

Richards, G. and Wilson, J. (eds) (2007) *Tourism, Creativity and Development. Contemporary Geographies of Leisure, Tourism and Mobility*, Vol. 10. Routledge, London.

Sattar, S. (2018) Culture, creativity and tourism: a study of Sikkim. *National Geographical Journal of India* 64(3–4), 241–248.

Sava, D. (2021) The creative tourism – an interactive type of cultural tourism. *Ovidius University Annals. Economic Sciences Series* 21(2), 486–492.

Scherf, K. (2021) Creative tourism in smaller communities: collaboration and cultural representation. In: Scherf, K. (ed.) *Creative Tourism in Smaller Communities: Place, Culture, and Local Representation*. University of Calgary Press, Calgary, Canada, pp. 1–26.

Smith, M. (2016) *Issues in Cultural Tourism Studies*. Routledge, New York.

Solène, P. (2011) Establishing the connections between the goals of sustainable development and creative tourism. Master thesis, Uppsala University, Uppsala, Sweden.

Souca, M.L. (2019) Revitalizing rural tourism through creative tourism: the role and importance of the local community. *Marketing – from Information to Decision Journal* 2(2), 43–50.

Sungsuwan, T. (2018) Creative tourism in Koh Samed: the answer for sustainable tourism. *NIDA Case Research Journal* 10(2), 97–118.

Sutarman, A., Rahardja, U., Putri Oganda, F., Millah, S. and Nur Azizah, N. (2023) The role of information technology in empowering the creative economy for sustainable tourism. *Aptisi Transactions on Technopreneurship (ATT)* 5(2sp), 175–185.

T.C. Kütahya Belediyesi (2023) Belediyemizden yeni bir proje, yeni bir hibe. Available at: https://www.kutahya.bel.tr/haber.asp?id=6633 (accessed 16 February 2024).

Teixeira, S.J. and Ferreira, J.J.M. (2019) Entrepreneurial artisan products as regional tourism competitiveness. *International Journal of Entrepreneurial Behavior & Research* 25(4), 652–673.

UNESCO (2006) Towards Sustainable Strategies for Creative Tourism. Discussion Report of the Planning Meeting for 2008 International Conference on Creative Tourism, Santa Fe, New Mexico, USA, October 25–27, 2006. Available at: https://unesdoc.unesco.org/ark:/48223/pf0000159811?posInSet=1&queryId=f6228c7f-62d1-4de9-9266-d70121deef8a (accessed February 2024).

UNESCO (2016) Türkiye'den Üç Unsur İnsanlığın Somut Olmayan Kültürel Mirası Temsili Listesine Kaydedil. Available at: https://www.unesco.org.tr/Home/AnnouncementDetail/142 (accessed 3 March 2024).

UNESCO (2024) Creative Cities Network - Kütahya. Available at: https://www.unesco.org/en/creative-cities/kutahya (accessed 12 January 2024).

Uzunget, E. (2019) *Seramik sanatçısı Mustafa Tunçalp ve eserlerinde çini sanatının etkileri. Yayımlanmamış Yüksek lisans Tezi*. Uşak Üniversitesi Sosyal Bilimler Enstitüsü, Uşak, Türkiye.

Zoğal, V. and Emekli, G. (2017) Yaratıcı turizme kavramsal ve coğrafi bir yaklaşım. *Ege Coğrafya Dergisi* 26(1), 21–34.

9 Digital Transformation and Rural Tourism: Rural SMTEs' Dynamics in Cyprus

Maria Hadjielia Drotarova* **and Vasilis Papavasiliou**
CTL Eurocollege, Limassol, Cyprus

Abstract

This chapter explores the workings of rural small and medium tourism enterprises (SMTEs) in Cyprus, with a focus on the incorporation of rural tourism and digital transformation. Rural tourism, an emerging niche, nurtures interactive and participatory experiences between tourists and local communities, fostering cultural and emotional exchanges. This form of tourism creates unforgettable experiences by engaging tourists with local culture, thereby contributing to the socio-economic progress of rural areas and sustainability. The chapter emphasizes the substantial challenges encountered by rural SMTEs, particularly in sustainability and innovation. Conventional tourism practices, such as sun-and-sea tourism, along with the repercussions of the COVID-19 pandemic, have heightened community deterioration and socio-economic hardships. The chapter accentuates the significance of digital transformation in addressing these challenges, encouraging innovation and promoting sustainable development. The findings divulge a deficiency in skills among rural SMTEs, constraining their capacity to make informed decisions, innovate and effectively utilize digital marketing. This deficiency jeopardizes their long-term sustainability and competitiveness. The chapter proposes that policy makers and industry stakeholders should prioritize interventions to enhance digital skills and support sustainable tourism development. By doing so, rural SMTEs can better harness local resources and contribute to the sociocultural and economic rejuvenation of their communities.

Keywords: Digital transformation, rural tourism, sustainability, SMTE, Cyprus

Introduction

For the past few decades, research on rural tourism has been crucial (Karali *et al.*, 2024; Trisia *et al.*, 2024) and many researchers have backed rural tourism as a way to support the rebirth of traditional villages. Indeed, traditional villages have been impacted by modernization and industrialization. They have experienced an ageing population, the migration of young highly educated individuals, a decline in vitality and the restructuring of rural industries (Karali *et al.*, 2024; Trisia *et al.*, 2024; Yanan *et al.*, 2024). As such, the need to revitalize the local economy and community in traditional villages is imperative and rural tourism has been acknowledged as a key developmental route for such regions. Rural tourism is recognized as a common approach to rural development in both developed and developing countries due to its potential to yield substantial social and economic benefits (Yanan *et al.*, 2024) and, hence, contribute to the sustainability of destinations. Even so, the sustainability of rural tourism has been questioned as it may disrupt local people's

*Corresponding author: hadjieliadrm@ctleuro.ac.cy

 Niche Tourism and Sustainability: Perspectives, Practices and Prospects (eds A. Farmaki *et al.*)
DOI: 10.1079/9781800626669.0009

lives, lead to overcrowding and damage the local environment (An and Alarcón, 2020). Evidently, the role of rural tourism businesses in fostering sustainability is crucial (Priatmoko *et al.*, 2023).

This chapter aims to explore the dynamics of rural small and medium tourism enterprises (SMTEs) in Cyprus. By researching the experiences, practices and challenges faced by these enterprises, the chapter seeks to offer a comprehensive understanding of their role within the tourism landscape. Furthermore, the chapter aims to identify the sustainability challenges encountered by rural SMTEs. The focus is to dive into the context of community degradation and socio-economic conditions influenced by traditional tourism practices. Traditional tourism practices include travel to well-known locations, participating in established tourism activities and participating in natural, historical and cultural attractions (Smith and Robinson, 2006). Doing so, the authors investigate the role of digital transformation in addressing the challenges identified that are faced by rural SMTEs. As such, the chapter offers insights that can inform practitioners of the strategies necessary for fostering sustainability and innovation within the rural tourism sector.

Theoretical Background

An increasing number of tourists seek to engage in active participation in their experiences and gain more meaningful contact with domestic people (Kastenholz and Gronau, 2022). These advancements, according to Della Lucia and Trunfio (2018), are essential in the planning of visitor experiences. Nowadays, there is much discussion on how to encourage the revitalization of rural tourism, especially in ways that promote sustainable tourism development (Richards, 2020). In this context, culture has been identified as a key element of the visitor experience, especially within rural settings (Ma *et al.*, 2021). Hence, culture-based rural tourism has taken a central role in these discussions as tourism has developed into a crucial tool for implementing innovative strategies into practice. A broader trend towards creating desirable destinations for people to live in, work in and travel to is reflected in the growth of rural tourism wherein creativity is used to reproduce cultural aspects (Richards, 2020; OECD, 2021). For instance, the advancement of innovation and digital transformation within rural tourism is observed including the promotion of investment in digital skills, particularly for tourism providers such as innovative producers and artisans (Ana, 2017; Boiko, 2020; Majdak and de Almeida, 2023). The sections below discuss the role of small and medium tourism providers in rural areas as well as their digital transformation attempts which have been argued to enhance the sustainability of rural tourism.

Rural tourism and SMTEs

Rural tourism has been considered as a new generation form of tourism, which can restart the tourism sector during and after the COVID-19 pandemic period (Vaishar and Šťastná, 2022). The first generation is beach tourism, in which people come to a place for relaxation and leisure, while the second is cultural tourism, oriented towards museums and cultural tours. Rural tourism is regarded as a third-generation tourism form that demands tourism stakeholders to evolve accordingly. They must, for instance, acknowledge the innovative potential of their rural communities as a resource and offer new approaches to cater to tourists' evolving needs (Aguiló *et al.*, 2005). Innovative and dynamic, sustainable and reliant on local resources, rural tourism development must be proactive and grounded in learning processes that are woven into the social fabric (Richards, 2011). In paving the way towards the advancement of innovation and in facilitating a shift towards rural tourism, we argue for a focus on SMTEs in rural areas. SMTEs play a significant role in promoting employment and sustainable economic growth in Europe and around the world (Ayyagari *et al.*, 2007).

Indeed, small and medium enterprises (SMEs) are frequently portrayed as 'job engines' and the backbone of economic growth (Mandl *et al.*, 2016). SMEs must adjust to a variety of internal and external factors. In addition to the typical difficulties faced by SMEs like resource constraints, SMEs are often significantly impacted and shaped by the environment in which they operate, such as whether they are situated in a rural or urban area. Particularly in a rural setting, SMEs can significantly contribute to economic and community growth and sustainable rural development (Hadjielias and Poutziouris, 2015; Hadjielias *et al.*, 2022). Compared with a more urban setting, owner-managers of rural SMEs are typically better ingrained in their communities and rural regions (Lang *et al.*, 2014; Korsgaard *et al.*, 2015; Zhao and Jones-Evans, 2017; Hadjielias *et al.*, 2022; Beckmann *et al.*, 2023). Indeed, SME owners' feeling of attachment to their regions is often higher within rural areas which in turn encourages rural SMEs to contribute actively to the sustainability of rural communities through new jobs, financial donations, infrastructures and local innovative initiatives (Lang *et al.*, 2014; Berlemann and Jahn, 2016; Hadjielias *et al.*, 2022). Unique historical, cultural or physical resources can also be found in rural areas (Korsgaard *et al.*, 2015), thereby creating opportunities for SMEs. Nevertheless, rural spatial contexts can also pose difficulties for SMEs, ranging from lower qualification levels and underdeveloped infrastructure to economic decline, underemployment and restricted export opportunities (Benneworth, 2004; Martin and Sunley, 2006; Horlings and Padt, 2013; Mayer and Baumgartner, 2014).

SMTEs in rural areas represent the majority of firms in the tourism sector (Villanueva-Álvaro *et al.*, 2017; Dimitriadou *et al.*, 2021). However, recent research highlights that these types of firms face serious limitations which undermine their sustainable growth (Saxena and Ilbery, 2008; Munar, 2012), the sustainability of rural communities (Hadjielias *et al.*, 2022) and tourism in rural areas (Ćurčić *et al.*, 2021; Ivona, 2021). SMTEs 'comprise all businesses, which by their own definition, operate in the tourism industry and employ up to 100 employees and include sole operators not employing any staff. Specifically, micro businesses are those that employ between one and four workers, small businesses employ between five and 19 employees and medium businesses employ between 20 and 100 employees' (Breen *et al.*, 2005, p. vi). The presence of SMTEs is a common feature of tourism destinations (Buhalis, 1996); in fact, these enterprises do well in the international tourism market (Buhalis, 1993). SMTEs provide a wide range of goods and services such as accommodation, food service, travel, entertainment and activities (Buhalis, 1996). Research on rural SMTEs reveals that the majority of them are known as 'gap-fillers', run by 'constrained' or 'non-entrepreneurs' who have not embraced any innovations in their products or processes (Ioannides and Petersen, 2003). Hence, in rural areas, SMTEs can mobilize their local resources, thus they could succeed in using and protecting their heritage (Petelca *et al.*, 2021).

SMTEs in rural areas through rural communities are the focus of several studies (Khokhobaia, 2019; Petelca *et al.*, 2021). Within SMTEs in rural areas, recent studies reveal that higher innovation levels are related to proactive and strong relationships with local actors (O'Connor, 2023). The actors who have a greater role in innovation are special customers, scientific organizations and non-human actors such as trade fairs (Ren, 2011). More diversity in proactive partnerships facilitates the development of higher-level innovations. Within SMTEs in rural areas, reactive and weaker relationships are typically associated with lower levels of innovation (Reidolf, 2016). Indeed, SMTEs in rural areas receive little assistance, organized investment policies or initiatives. A shared understanding of the opportunity, expertise and cooperation is required to guide their owners. SMTE owners in rural areas need to take more initiatives and work together with the government at all levels, thus to gain a more competitive position (Gregoric and Pajic, 2016) in the digital and sustainable international market.

In the context of rural tourism, nowadays tourists choose to look for SMTEs and engage in activities that offer exposure to cultural resources. Experiences with local cuisine and wine, traditions and rural activities as well as the authenticity of the rural setting are among the interests that are particular to one's self (Amaral, 2020; Maziliauske, 2024). Hence, considering the above, rural tourism as an innovative niche tourism form can be stimulated through the enhancement of digital innovation within SMTEs in rural areas. SMTEs in rural areas is a core group of tourism stakeholders under rural tourism, which need to be supported through relevant training and mentoring provision to help them overcome existing innovation and digital skill limitations that halt their digitalization potential. At the same time, SMTEs in rural areas can benefit from networking infrastructures and knowledge management systems that match them with local actors that are important in innovation such as governmental and non-governmental organizations, trade associations, financial corporations, centres for innovation and entrepreneurship, incubator-accelerator for startups, idea supporters and so on.

Digital transformation and SMTEs

The advent of digital transformation is effecting a profound transformation in the landscape of SMTEs (Mxunyelwa, 2016; Maingi and Wachira, 2022). This is occurring as a result of the ability of digital tools and technologies to enhance operational efficiency, customer engagement and competitive advantage. The implementation of digital tools and technologies including online booking systems, customer relationship management (CRM) software and social media marketing enables SMTEs to enhance operational efficiency, reduce costs and improve service delivery (Buhalis and Law, 2008; Mxunyelwa, 2016; Buhalis and Murphy, 2017; Maingi and Wachira, 2022). Furthermore, the utilization of data analytics enables these enterprises to gain insights into customer preferences and behaviour, thereby facilitating the implementation of more personalized and targeted marketing strategies. This transition enables SMTEs to reach a more extensive customer base while simultaneously developing more meaningful and engaging experiences for their users (Sigala, 2018; O'Connor, 2024).

The influence of digital transformation on SMTEs is not limited to operational enhancements; it also serves to stimulate innovation and resilience in the context of industry-specific challenges (Buhalis and Law, 2008). The utilization of digital platforms enables SMTEs to diversify their service offerings including the provision of virtual tours and digital concierge services, which have become particularly pertinent during the period of the global pandemic caused by COVID-19. Such innovations not only provide alternative revenue streams but also enhance the overall customer experience by offering greater flexibility and convenience (O'Connor, 2024). Furthermore, the implementation of digital transformation equips SMTEs with the requisite tools to adapt to the rapid changes observed in market conditions, thereby ensuring their sustainability and growth in a highly competitive and dynamic environment (Sigala, 2018; O'Connor, 2024). While local, national and European strategies promote digital transformation as a step towards building resilience and a more sustainable future (Sigala, 2018; Hernández Sánchez and Oskam, 2022; O'Connor, 2024), the study by Hernández Sánchez and Oskam (2022) identifies two main uncertainties that may jeopardize this transformation: (i) a change in the traditional 'sun, sand and sea' visitor to a more conscious, individual and curious 'promad'; and (ii) the corporate culture of SMTEs.

Methods

This chapter aims to explore the dynamics of SMTEs in rural areas in Cyprus, a popular tourist island in the Mediterranean. Specifically, it examines the impact of digital transformation and sustainability on rural SMTEs in Cyprus. By uncovering the subjective experiences of stakeholders, the study reported here contributes a deeper understanding of the challenges and opportunities faced by rural SMTEs in the context of sustainability and digital transformation in Cyprus. Thus, a qualitative research approach (Miles and Huberman, 1994) was deemed a natural fit as a methodology. This methodology is ideal since it deals with the understanding of human beliefs, behaviours, values and perceptions of particular social and non-social issues. As a method, the study pursued a phenomenological approach. The specific method provides, as suggested by Miles and Huberman (1994), a means to explore the essence of experiences. Hence, by using a phenomenological approach in investigation, rich descriptions and subjective insights can be obtained.

Given the above, an inductive research approach aligns perfectly with the research goals. As advocated by Braun *et al.* (2013), this approach allows the source of general principles to emerge from specific observations. In-depth interviews, a key component of the phenomenological approach, have been widely acknowledged as a valuable tool for capturing the depth of experiences (Marques and Richards, 2014). Consequently, the research methodology involved employed an inductive approach within the phenomenological framework, following the principles outlined by Miles and Huberman (1994). The study conducted in-depth interviews with 30 participants, primarily SMTE owner-managers in Cyprus, seeking to gain insights into the experiences of SMTEs in rural areas. Participants were purposefully selected to ensure a diverse representation of those directly involved in SMTEs and related activities. The data provided offers methodological grounding for this approach, emphasizing its applicability in exploring the multifaceted landscape of rural tourism.

Research questions

The research questions for this study were developed through a careful consideration of the intersection between rural tourism, sustainability, digital transformation and the specific context of rural SMTEs in Cyprus. The overarching aim was to address the challenges and opportunities faced by these enterprises within the tourism landscape, particularly in light of community degradation, socio-economic conditions influenced by traditional tourism practices and the COVID-19 pandemic. Drawing from the existing literature on rural tourism (Figueiredo, 2022), rural entrepreneurship (Gibson, 2012) and sustainable development (Horlings and Padt, 2013), the research questions were crafted to explore the dynamics of SMTEs' responses to sustainability challenges, the role of digital transformation in overcoming these challenges, the specific innovation and digital skill limitations they face and the potential contributions of their experiences to broader discussions on sustainable tourism development. By framing the research questions in this way, the study aimed to provide comprehensive insights into the complexities of rural tourism entrepreneurship and inform strategies for fostering sustainability and innovation within the sector. Consequently, the research questions that guided the study are as follows:

1. How do SMTEs in rural areas in Cyprus navigate and respond to sustainability challenges associated with community degradation and worsening socio-economic conditions, influenced by traditional 'beach and mass' tourism and the COVID-19 pandemic?
2. In what ways can digital transformation assist SMTEs in rural areas in Cyprus in overcoming the challenges posed by community degradation and socio-economic issues, and how can it contribute to the development of digital and green skills within the tourism sector?
3. What are the specific innovation and digital skill limitations faced by SMTEs in rural areas in Cyprus, and how do these limitations impact decision making, product development and marketing practices within the sector?
4. How can the experiences and practices of SMTEs in rural areas in Cyprus contribute to broader discussions on sustainable tourism development, innovation and digital transformation?

Data analysis

For qualitative data analysis a systematic approach that emphasizes the importance of rigour and structure in the inductive research process is important (Corley and Gioia, 2011; Gioia *et al.*, 2013; Gioia, 2021). Its robustness lies in the ability to ensure qualitative rigour through careful data handling and theoretical articulation. As outlined by Gioia *et al.* (2013), the methodology involves the creation of a data structure that visually represents the connections between first-order concepts and second-order themes, ultimately leading to the development of aggregate dimensions. This data structure not only provides a clear and transparent representation of the analytical process but also facilitates the construction of a coherent and compelling theoretical narrative.

According to Gioia (2021), our analytical process began with the identification of first-order concepts, which we took directly from the interview transcripts without any preconceived theoretical framework. These first-order concepts were then categorized into second-order themes through an iterative process of comparison and contrast, which allowed us to identify patterns and relationships within the data. This systematic approach ensured that the analysis remained grounded in the empirical data while moving progressively towards higher-level theoretical constructs.

Findings

In this section we present research findings from in-depth interviews with 30 owner-managers of rural SMTEs in Cyprus (see Table 9.1).

Table 9.1. Research participant profile.

No.	Pseudonym	Age (years)	Village
1	Andreas	45	Omodos
2	Maria	38	Omodos
3	Christos	50	Omodos
4	Eleni	42	Omodos
5	Giorgos	47	Laneia
6	Nicos	39	Laneia
7	Panayiota	55	Laneia
8	Yiannis	36	Laneia
9	Anna	44	Ayios Mamas
10	Kyriakos	49	Ayios Mamas
11	Dimitra	41	Ayios Mamas
12	Andreas P.	53	Ayios Mamas
13	Sophia	37	Trimiklini
14	Michalis	46	Trimiklini
15	Katerina	40	Trimiklini
16	Panicos	48	Trimiklini
17	Elena	39	Kakopetria
18	Stavros	52	Kakopetria
19	Fani	43	Kakopetria
20	Marios	35	Kakopetria
21	Marilena	38	Troodos
22	Christoforos	71	Troodos
23	Ioanna	45	Kalopanayiotis
24	Theodoros	50	Kalopanayiotis
25	Christina	42	Kalopanayiotis
26	Charalambos	47	Kalopanayiotis
27	Evi	39	Zygi
28	Andreas K.	54	Zygi
29	Agathi	44	Zygi
30	Petros	48	Zygi

Traditional Cypriot villages such as Omodos, Laneia, Ayios Mamas, Trimiklini, Kakopetria, Troodos, Kalopanagiotis and Zygi offer a unique range of tourism products and services that highlight their rich cultural heritage and natural beauty. Visitors can enjoy authentic experiences such as grape harvesting in Ayios Mamas, wine-tasting tours in Omodos, the Commandaria storehouse museum in Laneia, exploring ancient monasteries in Kalopanagiotis and indulging in traditional Cypriot cuisine in Zygi's famous fish tavernas. These villages also offer opportunities for nature walks in the Troodos Mountains, visits to historic sites and local craft shops, and participation in traditional festivals and cultural events.

The findings presented below revolved around three key themes that emerged from the analysis: (i) navigating sustainability and social challenges; (ii) digital transformation as a solution for SMTEs in rural areas; and (iii) innovation and digital skill limitations.

Navigating sustainability and social challenges

SMTEs in rural areas in Cyprus face significant sustainability challenges stemming from community degradation and worsening socio-economic conditions, influenced by traditional 'beach and mass' tourism and exacerbated by the COVID-19 pandemic. 'From Monday to Friday you do not find anybody here, all tourists are down at the seaside' (Christina, 42, Kalopanayiotis, glassmaker and souvenir shop owner). The research participants highlighted the need for 'more innovative approaches to sustainability to ensure the long-term viability' (Andreas, 45, Omodos, lacemaker, educator and souvenir shop owner) of their businesses. As Andreas expressed, 'I must innovate my live YouTube webinars on 'pipila' lace making. And not many tourists are interested to try it face-to-face'.

Even though SMTEs recognize the advantages of and need for innovative approaches, they argued that there is an issue in 'attracting and hiring employees because of their location and remoteness' (Christos, 50, Omodos, winery owner). In this respect, SMTEs face challenges associated with community degradation. Most younger employees prefer to live in the cities rather than in rural communities that lack entertainment facilities. They feel isolated and therefore rural communities face declining infrastructure and loss of cultural heritage. As Christos added: 'My daughter was working with me as an event manager, but she could not stay here longer. All her friends are in the city, she was here alone and it's not easy to come and go here every day'.

Consequently, because of the lack of staff and low profit generated by a low tourist flow and local attendance, SMTEs 'do not operate daily and at regular hours' (Andreas K., 54, Zygi, fish restaurant owner). This degradation impacts the 'attractiveness of rural areas as tourist destinations and affects the overall visitor experience' (Yiannis, 36, Laneia, souvenir shop owner). It is evident that Cyprus remains a 'sun and sea' destination as 'more tourists are interested in residing and exploring seaside areas' (Anna, 44, Ayios Mamas, coffee shop owner). This poses challenges for SMTEs operating in rural areas. 'Competition from larger tourism establishments, price pressures and changing consumer preferences' towards more authentic and sustainable experiences, said Panicos (48, Trimiklini, wine shop owner), affect the competitiveness of SMTEs.

Digital transformation as a solution

The research findings suggest that digital transformation can assist SMTEs in rural areas in overcoming the challenges posed by community degradation and socio-economic issues. 'We see that we have to learn digital skills and we welcome this change in our business. Otherwise, we cannot survive, and our children will not want to take over in the future' (Stavros and Fani, 52 and 43, Kakopetria, restaurant owners). The research participants identified digital tools and platforms as valuable resources for enhancing marketing efforts, improving customer engagement and streamlining business operations. As Sophia (37, Trimiklini, traditional pasta maker) expressed: 'I can sell more via my online shop abroad than from my shop here'. Additionally, digital transformation has the potential to contribute to the development of essential digital skills within the tourism sector.

Digital skill limitations

Following the arguments above, it was highlighted that there was difficulty in establishing 'wireless connection in some areas due to poor service provided by the telecommunications providers' (Elena, 39, Kakopetria, honey producer). Another issue raised was the 'lack of skill and interest by the older generation' (Agathi, 44, Zygi, souvenir shop owner) that does not seem to be invested in upgrading their SMTE. As Christoforos (71, Troodos, hotel and souvenir shop owner) expressed: 'We like to stay traditional. We do not believe in all this online stuff' and thus is constantly at odds with his daughter Marilena (38, Troodos).

It is evident that despite the potential benefits of digital transformation, SMTEs in rural areas in Cyprus face specific limitations in terms of innovation and digital skills. Limited access to technology, lack of digital literacy and financial constraints were identified as major barriers to adopting digital solutions. These limitations impact decision making, product development and marketing practices within the sector.

Discussion and Recommendations

The findings of the present study shed light on the challenges faced by SMTEs in rural areas in Cyprus, particularly regarding sustainability, digital transformation and innovation. Based on these findings, it is crucial to discuss recommendations and examples that can be implemented to address these challenges effectively. Implementing eco-friendly measures such as energy-efficient operations, waste reduction strategies and community engagement initiatives can help SMTEs promote environmental conservation and social responsibility (Bramwell and Lane, 2011). Additionally, diversifying tourism offerings to include cultural and nature-based experiences can attract responsible tourists seeking authentic and sustainable travel experiences (Giaoutzi and Nijkamp, 2006). However, it was expressed that there was a lack of skills and understanding by SMTEs in rural areas. It is recommended that the Cyprus Deputy Ministry of Tourism establish a precise plan specifically for SMTEs in rural areas and incorporate this plan and training within its National Tourism Strategy 2030 (Cyprus Deputy Ministry of Tourism, 2024). Barringer *et al.* (2005) demonstrated that employee training in SMTEs plays a critical role in maintaining growth and staying up to date with the latest trends and methods for improving customer service and generating profit.

Developing an online presence through websites, social media and online booking platforms can help SMTEs attract and engage with customers effectively (Buhalis and Law, 2008). The lack of exposure is a reason why SMTEs do not have the same customer reach as in larger companies (Andreou, 2019). Furthermore, investing in digital skills training programmes for SMTEs' owners and employees can empower them to utilize technology more efficiently and adapt to changing market demands (Gretzel *et al.*, 2015). Globalization is forcing many firms to adapt to new technological needs in order to survive. The participants expressed that although they recognize the advantages and need to incorporate new digital skills and practices in their businesses, the technical skills and know-how as well as applicability is lacking in rural areas. Hadjimanolis (1999) stated that innovation is a difficult undertaking for small businesses since they have limited resources and expertise. This finding seems to be true in Cyprus; hence, more light needs to be shed towards SMTEs in rural areas. It would, therefore, be beneficial for SMTEs to collaborate with local governments and industry associations to improve digital infrastructure and connectivity in rural areas, which can also facilitate the adoption of digital solutions by SMTEs (Sigala, 2015).

Despite the potential benefits of digital transformation, SMTEs in rural areas face specific limitations in terms of innovation and digital skills. To address these challenges, SMTEs can seek support from government initiatives, educational institutions and industry associations to provide digital literacy training and technical assistance. For example, establishing training programmes and workshops on digital marketing, e-commerce and online booking systems can help SMTEs improve their digital capabilities and competitiveness (Gretzel *et al.*, 2015). Moreover, fostering intergenerational knowledge transfer and mentorship programmes can encourage older generations to embrace digital technologies and overcome resistance to change (Hemmington and King, 2000).

Conclusion and Implications

In examining the impact of digital transformation and sustainability on rural SMTEs in Cyprus, the study reported in this chapter highlights a significant skills gap among SMTEs in rural Cyprus which in turn limits their decision making, innovation and digital marketing practices. Due to neglecting research, digital advancements and external expertise, these SMTEs jeopardize their long-term sustainability and creativity in tourism provision. The findings presented in this chapter underscore the need for policy makers and industry stakeholders to prioritize interventions fostering sustainable tourism development and digital innovation within rural communities.

The study findings confirm past studies suggesting that SMTEs in peripheral areas are limited in their decision making, innovation, digitalization and marketing practices (Ioannides and Petersen, 2003). Indeed, SMTEs in rural areas in Cyprus create new product/service plans and execute their business ideas without taking much research and development into consideration, ignoring the latest digital technological developments and without relying on external expertise (Gibson, 2012). There is little presence of long-term brand-building affinity and limited consideration of digital business models in their new services (Yaja and Kumar, 2021). Furthermore, SMTEs in rural areas are less frequently engaged in digital marketing compared with firms in urban or metropolitan areas. These limitations, particularly the low level of innovation and digital skills within SMTEs in rural areas and their owner-managers, can limit the creativity and sustainability of tourism provision in rural areas. However, a recent study conducted in Norway demonstrates that due to the contextual aspects of rurality, SMTEs in rural areas collaborate with a variety of local stakeholders to co-create innovative solutions for sustainability, supporting the sociocultural benefits of sustainable rural destinations through quality of life, inclusion, education, local sourcing and resource synergies (Maziliauske, 2024). As such, hopeful prospects are present in enhancing the sustainability of rural tourism through the co-creation of activities between SMTEs and other stakeholders in rural areas.

Overall, the findings presented in this chapter offer valuable insight for policy makers, industry stakeholders and researchers in relation to fostering sustainable tourism development and promoting digital innovation within rural communities. This chapter highlights the importance of collaborative efforts between stakeholders to address sustainability challenges and harness the potential of digital technologies in driving positive change within the tourism sector. This chapter also offers contributions to broader discussions as the experiences and practices of SMTEs in rural areas in Cyprus may provide valuable insights that can expand scholarship on sustainable tourism development, innovation and digital transformation.

References

Aguiló, E., Alegre, J. and Sard, M. (2005) The persistence of the sun and sand tourism model. *Tourism Management* 26(2), 219–231.

Andreou, E. (2019) Socio-economic challenges on building a strong online presence – case study small and micro businesses in Cyprus (in Greek). Available at: https://kypseli.ouc.ac.cy/bitstream/handle/11128/4369/%CE%9A%CE%A0%CE%A3-2019-00041.pdf (accessed 17 October 2024).

An, W. and Alarcón, S. (2020) How can rural tourism be sustainable? A systematic review. *Sustainability* 12(18), 7758.

Amaral, M. (2020) Creative tourism and tourist loyalty in rural tourism enterprises: A case study in Alentejo. In: Cabeça, S.M., Gonçalves, A.R., Marques, J.F. and Tavares, M. (eds) *Creative Tourism Dynamics: Connecting Travellers, Communities, Cultures and Places*. Grácio Editor, Coimbra, Portugal, pp. 29–48.

Ana, M.I. (2017) Ecotourism, agro-tourism and rural tourism in the European Union. *Cactus Tourism Journal* 15(2), 6–14.

Ayyagari, M., Beck, T. and Demirguc-Kunt, A. (2007) Small and medium enterprises across the globe. *Small Business Economics* 29(4), 415–434.

Barringer, B.R., Jones, F.F. and Neubaum, D.O. (2005) A quantitative content analysis of the characteristics of rapid-growth firms and their founders. *Journal of Business Venturing* 20(5), 663–687.

Beckmann, M., Garkisch, M. and Zeyen, A. (2023) Together we are strong? A systematic literature review on how SMEs use relation-based collaboration to operate in rural areas. *Journal of Small Business & Entrepreneurship* 35(4), 515–549.

Benneworth, P. (2004) In what sense 'regional development?': entrepreneurship, underdevelopment and strong tradition in the periphery. *Entrepreneurship & Regional Development* 16(6), 439–458.

Berlemann, M. and Jahn, V. (2016) Regional importance of Mittelstand firms and innovation performance. *Regional Studies* 50(11), 1819–1833.

Boiko, V.O. (2020) Green tourism as a perspective direction for rural entrepreneurship development. In: *Scientific Approaches to Modernizing the Economic System: Vector of Development: Collective Monograph*. Liha-Pres, Lviv, Ukraine/Toruń, Poland, pp. 1–18.

Bramwell, B. and Lane, B. (2011) Critical research on the governance of tourism and sustainability. *Journal of Sustainable Tourism* 19(4–5), 411–421.

Braun, E., Kavaratzis, M. and Zenker, S. (2013) My city – my brand: the different roles of residents in place branding. *Journal of Place Management and Development* 6(1), 18–28.

Breen, J.P., Bergin-Seers, S., Jago, L.K. and Carlsen, J. (2005) *Small and Medium Tourism Enterprises: The Identification of Good Practice*. CRC for Sustainable Tourism Pty Ltd, Gold Coast, Australia.

Buhalis, D. (1993) RICIRMS as a strategic tool for small and medium tourism enterprises. *Tourism Management* 14(5), 366–378.

Buhalis, D. (1996) Enhancing the competitiveness of small and medium sized tourism enterprises. *Electronic Markets* 6(1), 1–6.

Buhalis, D. and Law, R. (2008) Progress in information technology and tourism management: 20 years on and 10 years after the internet – the state of eTourism research. *Tourism Management* 29(4), 609–623.

Buhalis, D. and Murphy, H. (2017) Information communication technologies (ICTs), entrepreneurship and SMTEs. In: Ateljevic, J. and Page, S.J. (eds) *Tourism and Entrepreneurship: International Perspectives*. Routledge, London, pp. 287–300.

Corley, K.G. and Gioia, D.A. (2011) Building theory about theory building: what constitutes a theoretical contribution?. *Academy of Management Review* 36(1), 12–32.

Ćurčić, N., Mirković Svitlica, A., Brankov, J., Bjeljac, Ž., Pavlović, S. *et al*. (2021) The role of rural tourism in strengthening the sustainability of rural areas: the case of Zlakusa Village. *Sustainability* 13(12), 6747.

Cyprus Deputy Ministry of Tourism (2024) National Tourism Strategy 2023. Available at: https://www.visitcyprus.com/files/National_Tourism_Strategy_2030.pdf (accessed 17 October 2024).

Della Lucia, M. and Trunfio, M. (2018) The role of the private actor in cultural regeneration: hybridizing cultural heritage with creativity in the city. *Cities* 82, 35–44.

Dimitriadou, E., Bournaris, T., Stavrinoudis, T. and Iakovidou, O. (2021) The efficiency score of small accommodation businesses in non-coastal rural areas in Greece. *Sustainability* 13(19), 11005.

Figueiredo, A. (2022) What is creative tourism? *Travel Tomorrow*, 31 March 2022. Available at: https://traveltomorrow.com/what-is-creative-tourism/

Giaoutzi, M. and Nijkamp, P. (2006) *Tourism and Regional Development: New Pathways*. Ashgate Publishing, London.

Gibson, D. (2012) The cultural challenges faced by indigenous-owned small medium tourism enterprises (SMTEs) in Fiji: case studies from the Yasawa Islands. *The Journal of Pacific Studies* 32(2), 102–126.

Gioia, D. (2021) A systematic methodology for doing qualitative research. *The Journal of Applied Behavioral Science* 57(1), 20–29.

Gioia, D.A., Corley, K.G. and Hamilton, A.L. (2013) Seeking qualitative rigor in inductive research: notes on the Gioia methodology. *Organizational Research Methods* 16(1), 15–31.

Gregoric, M. and Pajic, A.K. (2016) The impact of small and medium entrepreneurship on tourism development in the Republic of Croatia. *Journal of Business Paradigms* 1(1), 85–94.

Gretzel, U., Sigala, M., Xiang, Z. and Koo, C. (2015) Smart tourism: foundations and developments. *Electronic Markets* 25(3), 179–188.

Hadjielias, E. and Poutziouris, P. (2015) On the conditions for the cooperative relations between family businesses: The role of trust. *International Journal of Entrepreneurial Behavior & Research* 21(6), 867–897.

Hadjielias, E., Christofi, M. and Tarba, S. (2022) Contextualizing small business resilience during the COVID-19 pandemic: evidence from small business owner-managers. *Small Business Economics* 59(4), 1351–1380. DOI: 10.1007/s11187-021-00588-0.

Hadjimanolis, A. (1999) Barriers to innovation for SMEs in a small less developed country (Cyprus). *Technovation* 19(9), 561–570.

Hemmington, N. and King, C. (2000) Key issues in the management of cultural heritage. *International Journal of Heritage Studies* 6(1), 3–6.

Hernández Sánchez, N. and Oskam, J. (2022) A 'new tourism cycle' on the Canary Islands: scenarios for digital transformation and resilience of small and medium tourism enterprises. *Journal of Tourism Futures*. DOI: 10.1108/JTF-04-2022-0132.

Horlings, I. and Padt, F. (2013) Leadership for sustainable regional development in rural areas: bridging personal and institutional aspects. *Sustainable Development* 21(6), 413–424.

Ioannides, D. and Petersen, T. (2003) Tourism 'non-entrepreneurship' in peripheral destinations: a case study of small and medium tourism enterprises on Bornholm, Denmark. *Tourism Geographies* 5(4), 408–435.

Ivona, A. (2021) Sustainability of rural tourism and promotion of local development. *Sustainability* 13(16), 8854.

Karali, A., Das, S. and Roy, H. (2024) Forty years of the rural tourism research: reviewing the trend, pattern and future agenda. *Tourism Recreation Research* 49(1), 173–200.

Kastenholz, E. and Gronau, W. (2022) Enhancing competences for co-creating appealing and meaningful cultural heritage experiences in tourism. *Journal of Hospitality & Tourism Research* 46(8), 1519–1544.

Khokhobaia, M. (2019) The role of small and medium size enterprises in regional tourism development: the case of Georgia. In: *Proceedings of the 2nd International Conference on Business, Management and Economics, Vienna, Austria, 21–23 June 2019* pp. 185–195. Available at: https://www.dpublication.com/wp-content/uploads/2019/06/64-974-ICBME24.pdf

Korsgaard, S., Müller, S. and Tanvig, H.W. (2015) Rural entrepreneurship or entrepreneurship in the rural – between place and space. *International Journal of Entrepreneurial Behavior & Research* 21(1), 5–26.

Ma, X., Wang, R., Dai, M. and Ou, Y. (2021) The influence of culture on the sustainable livelihoods of households in rural tourism destinations. *Journal of Sustainable Tourism* 29(8), 1235–1252.

Lang, R., Fink, M. and Kibler, E. (2014) Understanding place-based entrepreneurship in rural central Europe: a comparative institutional analysis. *International Small Business Journal* 32(2), 204–227.

Maingi, S.W. and Wachira, H.M. (2022) Digital skills and tourism workforce recovery in the post-COVID-19 pandemic era: case of small and medium-sized tourism enterprises (SMTEs) in Nairobi, Kenya. In: Korstanje, M.E., Seraphin, H. and Maingi, S.W. (eds) *Tourism Through Troubled Times: Challenges and Opportunities of the Tourism Industry in the 21st Century*. Emerald Publishing Limited, Bingley, UK, pp. 21–38.

Majdak, P. and de Almeida, A.M.M. (2023) UN sustainable development goals in the European periphery via rural tourism: Madeira. In: Filho, W.L., Dinis, M.A.P., Moggi, S., Price, E. and Hope, A. (eds) *SDGs in the European Region*. Springer, Cham, Switzerland, pp. 1475–1495.

Mandl, I., Hurley, J., Ledermaier, S. and Napierała, J. (2016) Job Creation in SMEs: ERM Annual Report 2015. Eurofound, Ireland. Available at: https://policycommons.net/artifacts/1841056/job-creation-in-smes/2583353/ (accessed 12 December 2023).

Marques, L. and Richards, G. (2014) The dimensions of art in place narrative. *Tourism Planning & Development* 4(2), 1–11.

Martin, R. and Sunley, P. (2006) Path dependence and regional economic evolution. *Journal of Economic Geography* 6(4), 395–437.

Mayer, H. and Baumgartner, D. (2014) The role of entrepreneurship and innovation in peripheral regions. *DisP – The Planning Review* 50(1), 16–23.

Maziliauske, E. (2024) Innovation for sustainability through co-creation by small and medium-sized tourism enterprises (SMEs): socio-cultural sustainability benefits to rural destinations. *Tourism Management Perspectives* 50, 101201.

Miles, M.B. and Huberman, A.M. (1994) *Qualitative Data Analysis: An Expanded Sourcebook*. SAGE Publications, Inc. Thousand Oaks, California.

Munar, A.M. (2012) Social media strategies and destination management. *Scandinavian Journal of Hospitality and Tourism* 12(2), 101–120.

Mxunyelwa, S. (2016) Information communication technology (ICT) as a management tool for small and medium tourism enterprises (SMTEs). *African Journal of Hospitality, Tourism and Leisure* 5(3), 1–15.

O'Connor, P. (2023) Small- and medium-sized tourism enterprises and smart tourism: tourism agenda 2030 perspective article. *Tourism Review* 78(2), 339–343.

O'Connor, P. (2024) Tourism enterprises and distribution: the effects of online food delivery platforms and online travel agencies. In: Hallak, R. and Lee, C. (eds) *Handbook of Tourism Entrepreneurship*. Edward Elgar Publishing, Cheltenham, UK, pp. 371–383.

OECD (2021) *Economic and Social Impact of Cultural and Creative Sectors: Note for Italy G20 Presidency Culture Working Group*. OECD Local Economic and Employment Development (LEED) Papers. OECD Publishing, Paris. DOI: 10.1787/4d4e760f-en.

Petelca, O., Pătrăuceanu, C.-C. and Beşliu, I. (2021) The multifunctionality and sustainability of small and medium enterprises in tourism-based rural communities. *The Annals of the University of Oradea* 30(2), 123–131.

Priatmoko, S., Kabil, M., Akaak, A., Lakner, Z., Gyuricza, C. *et al.* (2023) Understanding the complexity of rural tourism business: scholarly perspective. *Sustainability* 15(2), 1193.

Reidolf, M. (2016) Knowledge networks and the nature of knowledge relationships of innovative rural SMEs. *European Journal of Innovation Management* 19(3), 317–336.

Ren, C. (2011) Non-human agency, radical ontology and tourism realities. *Annals of Tourism Research* 38(3), 858–881. DOI: 10.1016/j.annals.2010.12.007.

Richards, G. (2011) Creativity and tourism: the state of the art. *Annals of Tourism Research* 38(4), 1225–1253.

Richards, G. (2020) Designing creative places: the role of creative tourism. *Annals of Tourism Research* 85, 102922. DOI: 10.1016/j.annals.2020.102922.

Saxena, G. and Ilbery, B. (2008) Integrated rural tourism: A border case study. *Annals of Tourism Research* 35(1), 233–254.

Sigala, M. (2015) Collaboration in tourism businesses and destinations: a systematic review. *Tourism Management* 51, 25–49.

Sigala, M. (2018) New technologies in tourism: from multi-disciplinary to anti-disciplinary advances and trajectories. *Tourism Management Perspectives* 25, 151–155.

Smith, M.K. and Robinson, M. (eds) (2006) *Cultural Tourism in a Changing World: Politics, Participation and (Re)presentation. Tourism and Cultural Change*, Vol. 7. Channel View Publications, Clevedon, UK.

Trisia, M.A., Prats, L. and Guia, J. (2024) Rethinking rurality in rural tourism: a new materialist approach. *Tourism Geographies*. DOI: 10.1080/14616688.2024.2360632.

Vaishar, A. and Šťastná, M. (2022) Impact of the COVID-19 pandemic on rural tourism in Czechia: Preliminary considerations. *Current Issues in Tourism* 25(2), 187–191.

Villanueva-Álvaro, J.J., Mondéjar-Jiménez, J. and Sáez-Martínez, F.J. (2017) Rural tourism: development, management and sustainability in rural establishments. *Sustainability* 9(5), 818.

Yaja, M. and Kumar, A. (2021) An empirical study of marketing of SMEs in the tourism sector. *Small Enterprise Research* 28(3), 314–328.

Yanan, L., Ismail, M.A. and Aminuddin, A. (2024) How has rural tourism influenced the sustainable development of traditional villages? A systematic literature review. *Heliyon* 10(4), e25627. DOI: 10.1016/j.heliyon.2024.e25627.

Zhao, T. and Jones-Evans, D. (2017) SMEs, banks and the spatial differentiation of access to finance. *Journal of Economic Geography* 17(4), 791–824.

Part IV: Prospects of Niche Tourism

This fourth and last part of the book consists of three chapters which consider the prospects of niche tourism in terms of destination sustainability. In the first chapter of the part (Chapter 10), Katerina Antoniou draws from the Israel and Palestine conflict to examine the role of citizen diplomacy that may be undertaken by residents and tourists in promoting peace tourism, which is argued to contribute to sustainable development. The next chapter (Chapter 11) by Gülsel Çiftci discusses the impact of cross-border tourism on social cohesion, justice and peace and, thereby, advances understanding of how this niche tourism form can contribute to sustainable development of destinations. The last chapter (Chapter 12), authored by Ângela Gonçalves, Jéssica Ribeiro, Mariana Oliveira and Bruno Barbosa de Sousa, offers an examination of volunteer tourism as a promoting sustainable niche tourism form that may exert positive impacts on local communities and support destinations' sustainability goals.

10 Citizen Diplomacy Through Peace Tourism

Katerina Antoniou*
School of Business and Management, University of Central Lancashire Cyprus, Larnaca, Cyprus

Abstract
Peace tourism is a niche form of tourist activity that, as its name suggests, is driven by an interest in peace. Peace tourism is a broad category of tourist activity that can have both positive and negative contributions to the host destination. This chapter explores the ways in which peace tourism enables individuals to satisfy this interest in peace by exercising citizen diplomacy. In doing so, it addresses a methodological gap in diplomacy research, which conventionally adopted a state-centric perspective and overlooked the political agency of the individual – and as such, the tourist. The chapter examines the ways in which citizen diplomacy is applied through peace tourism and whether it can positively contribute to sustainable peace within and beyond state boundaries. Using the case study of Israel/Palestine, this chapter identifies initiatives of citizen diplomacy through peace tourism and examines their impact using Anderson's model of six levels to peace. The study's findings show that citizen diplomacy through peace tourism can yield positive contributions to peace. The findings are discussed in reference to the United Nations' Sustainable Development Goals (SDGs) framework, and specifically SDG 16 on Peace, Justice and Strong Institutions.

Keywords: Citizen diplomacy, peace tourism, sustainable peace, Israel/Palestine, peace activism, SDG 16

Introduction

There is a growing scholarly interest in identifying the political impact of tourism and, more specifically, assessing the contribution of international tourist activity to sustainable peace. This growing area of research (Farmaki and Stergiou, 2021; Jamgade, 2021; Blanchard *et al.*, 2022; Antoniou, 2023) reconceptualizes tourist activity as one with direct political implications and, consequentially, redefines the tourist and the host as international political actors. The political activity of individuals has often been labelled as political activism and, more recently, as citizen diplomacy. The latter has been identified as a form of diplomatic activity that is 'by the people, for the people' and falls under the broader category of non-state diplomatic activity (Huijgh, 2016). Defining citizen diplomacy has been a contested process often featuring ambiguity, with the term being used to refer both to individuals as non-state political agents as well as individuals acting as representatives of state actors (Tyler and Beyerinck, 2016).

Redefining the individual as a distinct political actor and understanding how tourists and hosts engage in citizen diplomacy is a complex and challenging task, but also a key

*Corresponding author: kaantoniou@uclan.ac.uk

DOI: 10.1079/9781800626669.0010

step in advancing research on peace, specifically to inform the tourism and peace relationship. The challenge in acknowledging the tourist as a political agent delves from the convention that states have been treated as the primary actors of the international political stage and the ones to define diplomatic discourse and shape international relations. Methodologically, this has translated into a state-centric approach to the examination of international political discourse, often treating tourists as recipients and not drivers of political activity. Nevertheless, emerging research on citizen diplomacy has introduced individuals as separate political agents, acknowledging that citizens can deviate from their state's official narrative, foreign and public policy and engage in diplomatic activity through non-state channels, without becoming representatives of their state's position.

This realization is of particular importance for distinguishing between state-led and individually led forms of diplomatic activity. While the former contributes to identifying and measuring the impact and influence of state actors, the latter sheds light on an underexplored political agent, the individual. If individuals can engage in political and diplomatic activity that becomes impactful on an international scale and influences state-led political activity, then the individual's political agency could be catalytic for international diplomatic discourse. In reference to the peace and tourism relationship, exploring citizen diplomacy through tourism can offer valuable insights into tourism's contributions to sustainable peace. It is important to note that not all tourist activity can foster citizen diplomacy or engage with peace as a cause. International tourist activity represents a vast scope of activities, experiences and motivations, each yielding a distinct set of effects to the people and places involved. To reduce this spectrum of activity towards forms of tourism that are directly relevant to peace, this chapter considers peace tourism – in its various forms – as the type of tourist activity most relevant to the quest for peace. Defined as 'travel that is specifically motivated by and associated with conflict resolution practices and war prevention' (Antoniou, 2021, p. 323), peace tourism is effectively the amalgamation of tourist practices that actively seek to engage with, be informed about or foster peace. Peace tourism is, therefore, directly associated with acts of peace activism and citizen diplomacy, which makes peace tourism a niche tourism form that directly addresses sustainability – particularly sustainable peace.

Individuals in divided societies experiencing protracted conflict may seek to engage in political activity and peace activism beyond the state, particularly through peace tourism and citizen diplomacy. This chapter's case study, Israel/Palestine, offers an example of intractable conflict for which political leadership has failed to achieve any form of resolution, leaving citizens in prolonged division without indications of state-led reconciliation. Israel/Palestine has attracted a plethora of visitors for peace tourism, including educational visitors, activists, researchers and peace professionals. At the same time, citizens of protracted and intractable conflicts such as the example of Israel/Palestine have often sought alternative routes for connecting with members of the other community and forming alliances across the division to engage in political dialogue, or citizen diplomacy beyond state structures. Examples of citizen diplomacy across division lines have often been recorded to occur both beyond that state's territory, as well as through visitor experiences engaging international audiences locally, practices that both highlight the significance of travel in enabling citizen diplomacy for citizens of divided societies. To inform the tourism and peace relationship and reconceptualize the individual as a political actor, this chapter examines peace tourism as an enabler of citizen diplomacy in ways that contribute to sustainable peace. By employing the United Nations' Sustainable Development Goal 16 on Peace, Justice and Strong Institutions (SDG 16) as a point of reference, this chapter explores the ways in which citizen diplomacy through peace tourism fosters peace and whether these align with the notion of sustainable peace as determined through the SDG 16 targets and indicators.

Citizen Diplomacy: The Citizen as a Separate Political Agent

McDonald (2006) differentiates diplomacy between Track One and Track Two. Track One is the government-to-government diplomacy, which he refers to as 'fairly rigid, (...) not risk-taking, and (...) not very imaginative' (McDonald, 2006, p. 4). Track Two, on the other hand, is 'person to person, small group to small group, it's dynamic, it's risk-taking, it's imaginative, it gets things done' (McDonald, 2006). McDonald uses Track Two as a synonym for citizen diplomacy, which he then illustrates through a variety of activity categories – funding, activism, research, business, etc. – identified as a multitrack system, indicating the broader diversity of citizen diplomacy in comparison to state diplomacy.

Initiatives of citizen diplomacy have been linked to democratization as avenues of participatory democracy. The expansion of diplomatic discourse to engage individuals has been considered a democratization of diplomacy (Anton, 2022, 2023). For authoritarian regimes, citizen diplomacy has been seen as an effective means of accessing citizenry with restricted access to democratic processes and creating opportunities for public dialogue through workshops and conferences (Fulda, 2019). Geis *et al.* (2022) see citizen diplomacy as a product of a broader process of democratizing diplomacy, which enables citizen participation in processes of foreign policy. Nevertheless, considering citizen diplomacy as a form of participatory democracy and as a state-bound political activity restricts its potential to be seen as diplomatic activity beyond the state.

There is a distinction to be made between citizen diplomacy and public diplomacy. The latter is a state-led diplomatic activity which, according to Gilboa (2008), allows a state to engage with public opinion abroad and improve the state's image among foreign audiences. Public diplomacy is a form of diplomatic activity through which states or state-affiliated agencies engage with individuals from other territories with the aim of improving the state's image abroad (Melissen, 2013). It has also been treated as an umbrella term to incorporate various forms of diplomatic activity that enable states to engage with non-state actors and individuals – one such example being cultural diplomacy (Ang *et al.*, 2015). Citizen diplomacy can be diverse in an analogous manner to public diplomacy and utilize culture, education, the arts and other avenues for diplomatic discourse. The key differentiation between public and citizen diplomacy is that the former is a government-led activity, whereas the latter is citizen-led and occurs beyond the state. Fulda (2019) characterizes citizen diplomacy as *society-centric* and as a key component of the world's transition towards new forms of diplomacy.

In response to the diversity of non-state forms of diplomacy – including citizen-led diplomacy – Anton (2022) and Antoniou (2023) place citizen diplomacy under the umbrella term civil society diplomacy (CSD), which offers a society-centric alternative to state-centric diplomatic activity. Contrary to McDonald's (2006) argument that Track Two and citizen diplomacy refer to the same activity, Antoniou (2023) identifies Track Two, grassroots diplomacy and citizen diplomacy as three separate forms of CSD. Antoniou defines Track Two as diplomatic activity involving non-state actors, grassroots diplomacy as diplomatic exchange across non-state actors and individuals, and citizen diplomacy as diplomacy across individuals, or people to people.

While there is a general scholarly agreement on the nature of citizen diplomacy and its capacity to be citizen-led rather than state-led, there is an evident challenge in drawing the line between individuals as distinct political actors and individuals within organized and institutionalized units under the framework of civil society. Individuals may choose to advance their political activity through civil society groups, which become hubs for coordinated citizen diplomacy. Such groups can vary from non-formal partnerships of a few individuals to transnational civil society movements engaging hundreds of thousands.

A common characteristic of the various forms of citizen diplomacy, and one that differentiates them from conventional, state-led diplomacy, is that they are non-formal in nature and there is no formal process or conditions under which an individual's activity can be classified as diplomatic discourse. This non-formal nature of citizen diplomacy can be seen as the principal factor of why it was not considered as an integral component of international diplomacy until recently, when globalization and increased mobility have given individuals more opportunities for political engagement beyond their state.

Anton and Moise (2022) consider an individual's global recognition and ability to exert global influence as a factor that elevates their capacity as a citizen diplomat and enables them to have global reach and diplomatic power. Global figures such as Angelina Jolie, Leonardo DiCaprio and Emma Watson, among others, are figures with global outreach who have used their position to speak out about global causes such as peace, climate action and women's rights. While internationally known figures have an established platform for conducting citizen diplomacy, a plethora of both community-based and transnational civil society networks provide a platform to individuals regardless of their level of global influence to serve global causes and conduct citizen diplomacy.

Citizen diplomacy for peace

Citizen diplomacy has been heralded as a means enabling individuals to serve peace. Yaniv (2013) discusses citizen diplomacy in reference to the Israeli–Palestinian conflict and highlights the importance of citizen involvement in reconciliation discussions. Fulda (2019) considers citizen diplomacy to be an alternative problem-solving route to problematic interstate relations, which are often examined solely on a government-to-government level. A prerequisite for examining the contribution of citizen diplomacy to peace is to define peace and identify suitable indicators for its measurement. Anderson (2004) defines peace as a condition characterized by harmonious relationships and low levels of violence and suggests that there are various levels for achieving it from a micro to macro scale. The seven levels presented are the personal level, the interpersonal, the intercultural, the intercommunal, the national, the international and the ecological (Anderson, 2004). This definition highlights the need to look beyond the state level for defining and measuring peace and acknowledge key areas of peace that have remained largely underexplored by state-centric research, such as the interpersonal, the intercultural and the intercommunal. Isaac (2014) agrees that peace does not refer solely to the relationships between nations, but should be extended to incorporate communities, groups and individuals. These less observed levels are where citizen diplomacy takes place more evidently.

A prominent example of citizen diplomacy for peace is the personal initiative and advocacy demonstrated by Nobel Peace Prize winners, who were awarded for their ability to establish routes for diplomacy beyond their state in cases where human rights violations were performed. 2023 Nobel Laureate Narges Mohammadi received the Nobel Peace Prize for advocating for women's rights and fighting against women's oppression under Iran's theocratic regime (Nobel Prize, 2024). In 2021, the Nobel Peace Prize was awarded to two investigative journalists, Maria Ressa and Dmitry Muratov, for their fight to secure freedom of expression and press freedom. Ressa and Muratov, active in the Philippines and Russia, respectively, were publicly critical of their governments' actions. Their activity towards safeguarding freedom of expression was heralded as a precondition to democracy and sustainable peace (Nobel Prize, 2024). In 2014, the Nobel Peace Prize was shared between Kailash Satyarthi and Malala Yousafzai for their actions in India and Pakistan, respectively, to protect the right of children to access education.

There is a notable observation to be made regarding citizen diplomacy and its impact. Although the activity is seen as citizen-led, the overwhelming share of citizen diplomacy cases discuss activity that individuals take in reference to their state. In the case of the five Nobel Laureates, their contributions to sustainable peace were made within their state boundaries and in reference to challenges caused by their state governments. Anton (2023) highlights that citizen diplomacy, like most forms of diplomacy, is examined from a state-centric approach, which restricts efforts for conceptualizing citizen diplomacy as a separate international political activity. It therefore seems that citizen diplomacy research has expanded more as a people-to-state phenomenon and less in its capacity as a people-to-people activity. When research is primarily concerned with how citizen diplomacy serves states, it expectedly ignores the individual's agency, organizational mobility and autonomy, and as a result fails to assess the political impact of individuals as separate political actors.

To this end, there is a methodological gap to be addressed in the study of diplomacy, which is the failure to assess individuals as political agents within the international political stage.

The inability to methodologically acknowledge individuals as political actors is caused by the state-centric approach that has overwhelmingly been applied when discussing and analysing diplomatic discourse. Mingst *et al.* (2018) identify the state-centric approach as one of three levels of analysis employed in the field of international relations. Namely, the three analytical approaches are: (i) the state level of analysis, for which the state is the principal actor; (ii) the individual level, which focuses on individuals and political interactions at a micro level; and (iii) the systemic level, which puts more emphasis on the system's structure at a macro level rather than examining its agents (Temby, 2015). Addressing the gap caused by state-centric approaches requires moving beyond the state as the principal political actor. For the purposes of this chapter's analysis, the individual level is deemed as the most appropriate for discussing the political impact of individuals as separate political agents. Shifting the focus to the individual as a political agent who is separate from their state's decision-making processes allows us to understand the evident divergence between state and citizen agency, and therefore formulate a clearer understanding of international political dynamics and contributions to sustainable peace.

Citizen diplomacy through peace tourism: Active beyond state boundaries

In a state-dominated international stage, political activity often follows a linear Actor (states)–Process (political activity)–Impact (individuals) relationship. From the state perspective, political decisions can have an impact on various individual activities, such as travel and tourism. A prominent example is the recent COVID-19 restrictions that states imposed on their citizens and territory, restricting or preventing travel and mobility, with a profound impact on individuals. States deciding to endorse international treaties, to wage war, to impose embargos or to revisit geopolitical alliances are all examples of state-driven international political affairs that affirm the above linear relationship and see individuals as recipients of political discourse and not as political agents.

Nevertheless, this chapter focuses on the individual's political agency – exercised through citizen diplomacy – and examines the reversed relationship Actor (individuals)–Process (political activity)–Impact (international political stage). Moving beyond the state perspective, this approach focuses on how individuals, as separate political agents, engage in political activity that can influence other political actors and the international political stage as a whole. Examining citizen diplomacy enables this shift in perspective, which encapsulates political activity driven by the individual as a political actor (Antoniou, 2023).

Anton (2023) acknowledges that individuals as citizen diplomats have separate political agency and do not act as representatives of their states. By using the term *unattached diplomat*, Anton (2023) emphasizes that, unlike traditional diplomats, individuals should not only be seen as instruments for state action nor as mere recipients of state-driven discourse, but as agents. Arguably, key factors for achieving separate agency for the unattached diplomat are transnational communication and mobility.

To this end, this chapter examines the contribution of individuals as citizen diplomats to causes beyond their national boundaries, such as sustainable peace, with a focus on transnational citizen diplomacy for peace. While transnational citizen diplomacy is a novel concept, it has in fact been practised for decades. The notion of global civil society is an example that highlights citizen action on a transnational scale, in the form of global activism (Kaldor, 2003, 2020; Keane, 2003). With globalization having enabled global governance, transnational communication and the emergence of non-state actors, political activity does not take place on a territorial basis, but rather in reference to specific issues (Kaldor, 2020). Moving beyond state boundaries has allowed individuals to tackle global issues through coordinated action that is informed by a variety of perspectives.

Noy (2013) discusses how peace activism can be achieved through tourism, not from the visitor's perspective, but through local initiative. Noy speaks about locals offering tourist experiences and narratives that act as alternatives to hegemonic narratives and aim to raise visitor consciousness. Using examples from Israel and Palestine, Noy illustrates how peace activism through tourism can challenge the apolitical nature of specific locations and question the way national ideology is enacted through tourism. Wintersteiner and Wohlmuther (2014) identify three types of peace-sensitive tourism, or tourism through which participants assume a responsibility to promote a culture of peace. These are: (1) tourism as an experience of the other; (ii) cross-border tourism and reconciliation tourism; and (iii) peace tourism.

Peace tourism, or tourism about peace and for peace as Wintersteiner and Wohlmuther define it, involves a niche audience of tourists with a direct interest in peace issues and enables them to advance their peace work through new knowledge, exchange of perspectives and dialogue (Wintersteiner and Wohlmuther, 2014). Antoniou (2022) explains peace tourism as 'a type of special interest tourism generated by the traveller's interest in peace', emphasizing that this interest alone cannot guarantee that peace tourism will have a positive contribution to peace. Instead, to assess the contributions of peace tourism to peace, Antoniou identifies a spectrum of peace tourism activity, which differentiates peace tourism that is informed and intentional from passive or unintentional peace tourism that could potentially lead to encounters counterproductive to the peace cause.

Moufakkir (2010) considers peace tourism to have emerged as the antithesis of political tourism, which is utilized to promote nationalist and divisive narratives, particularly in contested and divided destinations. Moufakkir's position aligns with Noy's (2013) argument that peace-motivated tourism can promote alternative narratives and foster peace activism, or, more broadly, citizen diplomacy for peace. Peace tourism can therefore offer peace-oriented mobility for visitors interested to actively pursue peace as a cause beyond national borders and potentially link them to local audiences offering peace tourism experiences as alternatives to either contested and divisive political narratives or apolitical versions of contested territories. Peace tourism, therefore, serves as an avenue for citizen diplomacy linking members of host and visiting territories under a common cause of intentionally serving peace.

Albeit a growing volume of literature discussing peace tourism and its potential over the past decades, the contribution of citizen diplomacy to peace through peace tourism has remained largely underexplored. There is, therefore, a scholarly gap to be addressed by shedding light on acts of citizen diplomacy through peace tourism. In addition to the methodological gap identified for examining citizen diplomacy beyond state-centric approaches, this chapter seeks to address the following two questions:

1. Can individuals utilize peace tourism to practise citizen diplomacy and make significant contributions to peace beyond their state affiliation?
2. Is it possible to identify the contribution of citizen diplomacy to peace using a non-state-centric methodology?

Methodology

To meet its research objectives, this chapter adopts an international relations analytical model to assess political activity and impact. From a selection of three levels of analysis, the state level, the individual level and the systemic level, this chapter moves beyond state-centric political analysis previously applied to diplomatic discourse and adopts the individual level of analysis for assessing citizen diplomacy conducted through peace tourism.

Through a review of secondary sources, the study identifies empirical data on citizen diplomacy initiatives through peace tourism to and from the destination of Israel/Palestine. A case study that has extensively contributed to research on peace tourism, peace activism and activist tourism, Israel/Palestine can offer valuable insights on citizen diplomacy and the ability of individuals to foster peace as distinct political agents. Through a review of peace tourism activities as recorded in relevant scholarship, the study identifies cases of citizen diplomacy conducted through peace tourism. These cases are then discussed in reference to their contribution to peace, using both Anderson's (2004) classification of seven levels of peace, as well as the sustainable peace indicators identified through SDG 16 on Peace, Justice and Strong Institutions.

Findings and Discussion

People-to-people diplomatic initiatives have become an integral component of peace work in the case of Israel/Palestine, with Israelis and Palestinians finding non-state avenues for common action (Isaac, 2014). Table 10.1 identifies organized groups and citizen initiatives in Israel/Palestine that engage with peace work, exercising citizen diplomacy through a variety of host and visitor peace tourism activities.

The initiatives listed above have utilized peace tourism through two approaches: (i) travelling abroad to conduct citizen diplomacy through peace tourism; and (ii) establishing community-based structures within the host destination to engage with peace tourists – domestic and international. This observation highlights that peace tourism is an avenue for citizen diplomacy available to both host and visitor audiences, who utilize it in a variety of forms to advocate for peace.

There is a common element in all the initiatives of citizen diplomacy through peace tourism that are identified in this case study, both for initiatives conducted within the destination as well as abroad. This common element is the initiatives' attempt to provide access to alternative, peace-oriented, under-represented and often marginalized narratives. In the case of Israel/Palestine, initiatives by local citizen diplomats have exposed international and domestic audiences to multifaceted narratives, contested testimonies, recordings, guided walks and unlikely partnerships in a way that promotes empathy, challenges nationalist and divisive rhetoric, humanizes the local communities and establishes opportunities for peace action.

Participants of Mejdi's dual-narrative tours have reported a reduction in prejudice (Schneider, 2019), while Jewish diaspora participants of Walls2Windows reported increased sympathy towards the Palestinian community (Schneider, 2020). Abed and Green, Standing Together and Women Wage Peace have engaged diplomats and policy makers within and beyond Israel to convey their message to state actors as well. Women Wage Peace and Women of the Sun have significantly expanded their global reach, joining events such as the Bled Strategic Forum and the Time Magazine 2024 Women of the Year ceremony (Women Wage Peace, 2024b).

From an admittedly wide plethora of local citizen diplomacy initiatives, including several initiatives on human rights and the reporting of military violence that were not included in Table 10.1, more than half have utilized tours, guided walks and interaction with visitor audiences to promote peace activism and offer insights into the destination's peace prospects in a people-to-people manner. Participants in the community-based initiatives for citizen diplomacy reported, in their majority, increased empathy towards both Israelis and Palestinians, as well as significant change in their understanding of the Israeli–Palestinian conflict (Schneider, 2019, 2020; Esu, 2021).

Observations about the impact of these initiatives are discussed from an individual level of analysis in reference to Anderson's (2004) levels of peace listed below:

1. Personal, inner peace.
2. Interpersonal peace within the group.
3. Intercultural peace among social groups.
4. Local peace within the community.
5. National peace within the nation.
6. International peace among nations.
7. Ecological peace with the Earth.

On a personal level (Level 1), significant changes in perception were recorded, and at the same time, participants were exposed to peace within groups of unlikely collaborators (Level 2), to peace among social groups (Level 3) and to peace within the visited communities (Level 4). There is less impact recorded on a national and ecological level; however, exposure to alternative narratives arguably affects the international level (Level 6) by reaching international audiences. Looking at these initiatives from a state-centric approach, one could argue that there is negligible if any impact. However, when delving into the levels of peace experienced on a micro level from an individual level of analysis, it is evident that citizen diplomacy through peace tourism can have a transformational effect on individuals and enable them to reconceptualize their understanding of divided societies and their dynamics in a post-state manner.

Methodologically, the non-formal nature of citizen diplomacy makes it difficult to examine

Table 10.1. Peace tourism activities in Israel/Palestine.

	Organization/initiative	Description	Peace tourism host activity	Peace tourism visitor activity	Source
1.	Sala Manca	Jerusalemite activist group wishing to expose the hegemonic politics of tourism in Jerusalem	Challenges nationalistic narratives through guided walk recordings Engages with peace activism through art projects and exhibitions	N/A	Noy (2013)
2.	Emek Shaveh	Israeli non-profit association of archaeologists and human rights activists (established 2008)	Countering the phenomenon of abusing archaeological sites to prove ownership of the land by offering alternative archaeological tours in East Jerusalem	N/A	Noy (2013)
3.	Mejdi	Tour operator	Delivers dual-narrative tours in Israel/Palestine	N/A	Schneider (2019)
4.	Israel/Palestine Center for Research and Information (IPCRI)	Public policy think tank	Delivers the Breaking Down Wall tours	N/A	Belhassen (2023)
5.	Ir Amim	Israeli organization	Delivers the United Jerusalem tour	N/A	Belhassen (2023)
6.	Roots/Shomashim/Judur	Israeli–Palestinian grassroots initiative	Visitors can access the West Bank Palestinian and West Bank Israeli settler members of the initiative in a central location of the West Bank and discuss peace and collaboration	N/A	Belhassen (2023)

7.	Breaking the Silence	Veteran Israeli soldier organization	Delivers tours to offer a dissonant voice to the narrative of militarism and provide personal testimonies to challenge the status of military occupation	N/A	Esu (2021)
8.	Walls2Windows	Tour programme	Delivers guided visits to the Occupied Palestinian Territories to the Jewish diaspora (mainly Jewish Americans)	N/A	Schneider (2020)
9.	Zochrot	Tour operator (established 2002)	Delivers tours to depopulated Palestinian sites to challenge national-historical narratives, promote awareness on Palestinian localities, and foster dialogue and reconciliation between Jews and Palestinians	N/A	Aviv (2011); Noy (2013); Belhassen (2023)
10.	Encounter	Tour operator	To challenge national-historical narratives, and promote dialogue and reconciliation between Jews and Palestinians	N/A	Aviv (2011)

Continued

Table 10.1. Continued

	Organization/initiative	Description	Peace tourism host activity	Peace tourism visitor activity	Source
11.	Women Wage Peace	Israeli women's organization (established 2014)	Hosts local peace activism events to promote women leadership in Israel across political spectrums for a political agreement in the conflict; to establish intercultural dialogue and joint action with Palestinian women	Travels internationally to promote the organization's message in partnership with their Palestinian sister organization	Bartolini de Angeli (2023); Women Wage Peace (2024a)
12.	Women of the Sun	Palestinian women's organization	Hosts local peace activism events to promote women's involvement in leadership for peace, and to establish intercultural dialogue and joint action with Israeli women towards peace Delivers youth camps for international audiences	Travels internationally to promote the organization's message in partnership with their Israeli sister organization	Women Wage Peace (2024a); Women of the Sun (2024)
13.	The Mothers' Call	An initiative run by the partnership between Women Wage Peace and Women of the Sun to promote cross-border motherhood for peace	Initiative promoted through local events and demonstrations	Initiative communicated abroad in various forums	Women Wage Peace (2024a)

14.	Tourism4Peace Forum	Initiative by the Israel Hotel Managers Association and the Peres Center for Peace to bring together tourism and hospitality providers from Israel, Egypt, Jordan and Palestine for peaceful, coordinated action (established 2005)	Promotes opportunities for peace tourism as an alternative to political tourism and exposes visitors to multiple narratives from all parties involved	N/A	Moufakkir (2010)
15.	Sally Abed Alon-Lee Green	Citizen initiative by Israeli Jewish and Israeli Palestinian individuals to tour university campuses in the USA and discuss peace for Israel/Palestine	N/A	Deliver talks to university students at Harvard and other campuses in the USA to advocate for unity and peace	Boorstein (2023)
16.	Standing Together	Grassroots Arab–Jewish peace advocating group in Israel	N/A	Trips to the USA to advocate for peace for Israel/Palestine	Boorstein (2023)

N/A, not applicable.

whether any of these initiatives caused a multiplier effect with participants further engaging in citizen diplomacy for peace as separate political agents. It is therefore challenging to explore the relationship between peace tourism participants and the international or ecological levels of peace in detail. This area of citizen diplomacy research has the potential to be further informed through empirical data that examine the long-term attitudes of peace tourism participants, both hosts and visitors.

With reference to SDG 16, the initiatives of citizen diplomacy through peace tourism identified in this case study are more closely relevant to target 16.10, which aims to 'ensure public access to information and protect fundamental freedoms, in accordance with national legislation and international agreements' (United Nations, 2024). By providing access to alternative narratives, challenging divisive and nationalist rhetoric, and offering the opportunity for dialogue across local and international audiences, these initiatives directly contribute to sustainable peace through one of its key targets. Nevertheless, SDG 16 targets and indicators reflect at large the national and international levels of peace, and do not sufficiently monitor peace work done at the micro levels of personal, interpersonal and intercultural peace. Advancing citizen diplomacy research to measure impact on the macro levels of peace (Levels 5, 6 and 7), and at the same time expanding sustainable peace targets to grasp peace activity at the micro level (Levels 1, 2, 3 and 4), can achieve a more comprehensive understanding of citizen diplomacy and peace tourism's contribution to peace.

Conclusion

This chapter sought to examine the political impact of individuals as non-state actors in reference to SDG 16 and sustainable peace. More specifically, the study focused on the political impact that can be generated through tourism-enabled citizen diplomacy. The relationship between tourism and peace has been extensively explored to assess whether there is significant causality of the former to the latter. In responding to this quest, Farmaki (2017) and Antoniou (2023) highlight that the question to be asked is not whether tourism leads to peace, but rather which forms of tourism have the capacity to do so, and under which circumstances.

Identifying the niche activity of peace tourism is a key step towards answering this question and effectively informing the tourism and peace relationship. Peace tourism is the collection of tourist activities that are motivated by an interest in peace and are, therefore, a principal reference point for assessing the impact tourism can have on peace: peace tourism activities have the potential to be intentional, informed and destination-driven – with host and visitors utilizing peace tourism to conduct citizen diplomacy as independent political agents – either by hosting international peace tourist audiences or visiting destinations abroad as peace tourists.

This chapter delved into the impact of peace tourism by discussing its ability to foster citizen diplomacy beyond the state. Through a review of peace tourism examples that involve travel to or from the case study of Israel/Palestine, it is evident that peace tourism has the capacity to encourage political activity through citizen diplomacy and create alternative routes for political action beyond state channels and state diplomacy. The activities reviewed indicate that peace tourism can foster citizen diplomacy initiatives and generate a positive impact to sustainable peace on the personal, interpersonal, intercultural, local/communal and international levels.

A key observation that emerges from this research is that moving beyond state-centric perspectives in diplomacy and peace research is imperative in understanding emergent forms of diplomatic discourse such as citizen diplomacy and in acknowledging the impact of individuals as political agents. Adopting the individual level of analysis addresses this methodological gap and informs diplomacy and peace from an individual perspective that more fittingly records tourist activity. Future research should delve further into the political impact of individuals on sustainable peace through their capacities as hosts and visitors.

Introducing niche tourism as an enabler of international political activity informs the fields of both tourism and international relations by identifying the political impact of tourism and by acknowledging niche tourists as actors within the international political stage. At the same time, a focus on the political impact of peace tourism provides direct theoretical implications to the study of sustainable peace and allows researchers to expand on the tourism and peace relationship from novel perspectives.

On a practical level, peace tourism has evolved to offer meaningful intergroup contact and experiences that significantly transform host and visitor perspectives. The insights on the transformational capacity of peace tourism recorded through this chapter's case study of Israel/Palestine make it a prominent peacebuilding tool for rapprochement and reconciliation in divided societies. When it comes to protracted conflicts, peace tourism becomes catalytic in providing access to marginalized, peace-oriented narratives and fostering intergroup connections within and beyond state boundaries, making it a valuable tool for achieving sustainable peace. To this end, the political contribution of niche tourism and, more specifically, the contribution of peace tourism to sustainable peace ought to be further examined within the realm of diplomacy studies and international relations.

This chapter has provided insights that acknowledge the individual as a non-state political actor with agency within the international political stage. If peace tourism has the capacity to endorse and amplify the political agency of individuals beyond the state, then state-owned deadlocks in protracted conflicts and divided societies can be challenged through novel avenues of citizen diplomacy.

References

Anderson, R. (2004) A definition of peace. *Peace and Conflict: Journal of Peace Psychology* 10(2), 101–116.

Ang, I., Isar, Y.R. and Mar, P. (2015) Cultural diplomacy: Beyond the national interest. *International Journal of Cultural Policy* 21(4), 365–381.

Anton, A. (2022) Conceptual pathways to civil society diplomacy. In: Sebastião, S.P. and Carvalho Spínola, S. (eds) *Diplomacy, Organisations and Citizens: A European Communication Perspective*. Springer, Cham, Switzerland, pp. 81–98.

Anton, A. (2023) Profiling a niche actor of civil society diplomacy: The unattached diplomat. *Journal of Communication Management* 27(2), 191–206.

Anton, A. and Moise, R. (2022) The citizen diplomats and their pathway to diplomatic power. In: Sebastião, S.P. and Carvalho Spínola, S. (eds) *Diplomacy, Organisations and Citizens: A European Communication Perspective*. Springer, Cham, Switzerland, pp. 219–254.

Antoniou, K. (2021) The role of visiting professionals in peacebuilding. In: da Silva, J.T., Breda, Z. and Carbone, F. (eds) *Role and Impact of Tourism in Peacebuilding and Conflict Transformation*. IGI Global, Hershey, Pennsylvania, pp. 320–341.

Antoniou, K. (2022) Peace tourism. In: Buhalis, D. (ed.) *Encyclopedia of Tourism Management and Marketing*. Edward Elgar Publishing, Cheltenham, UK, pp. 448–451.

Antoniou, K. (2023) *Tourism as a Form of International Relations: Insights from Contemporary Practice*. Edward Elgar Publishing, Cheltenham, UK.

Aviv, C. (2011) The emergence of alternative Jewish tourism. *European Review of History* 18(1), 33–43.

Bartolini de Angeli, E.L. (2023) The contribution of women to peace in the Middle East: The experience of the movement women wage peace (WWP). *Religions* 14(7), 820.

Belhassen, Y. (2023) Tourism as a space of possibilities: Reconciliation tourism enterprises in a conflict-ridden destination. *Journal of Travel Research* 62(7), 1497–1509.

Blanchard, L.-A., Hatakeyama, S. and Kawasaki, A. (2022) Making waves: Peace boat Japan as a model of sustainable peace through tourism. In: Higgins-Desbiolles, F., Blanchard, L.-A. and Urbain, Y. (eds) *Peace Through Tourism: Critical Reflections on the Intersections between Peace, Justice and Sustainable Development*. Routledge, London, pp. 277–294.

Boorstein, M. (2023) Israeli peace activists tour U.S. campuses and find a 'new war zone'. *The Washington Post*, 17 November 2023.

Esu, A. (2021) Voicing the silence: The naturalisation of violence under the rule occupation. *International Review of Sociology* 31(1), 159–181.

Farmaki, A. (2017) The tourism and peace nexus. *Tourism Management* 59, 528–540.
Farmaki, A. and Stergiou, D. (2021) Peace and tourism: Bridging the gap through justice. *Peace & Change* 46(3), 286–309.
Fulda, A. (2019) The emergence of citizen diplomacy in European Union–China relations: Principles, pillars, pioneers, paradoxes. *Diplomacy & Statecraft* 30(1), 188–216.
Geis, A., Opitz, C. and Pfeifer, H. (2022) Recasting the role of citizens in diplomacy and foreign policy: Preliminary insights and a new research agenda. *The Hague Journal of Diplomacy* 17(4), 614–627.
Gilboa, E. (2008) Searching for a theory of public diplomacy. *The Annals of the American Academy of Political and Social Science* 616(1), 55–77.
Huijgh, E. (2016) Public diplomacy. In: Constantinou, C.M., Kerr, P. and Sharp, P. (eds) *The SAGE Handbook of Diplomacy*. SAGE Publications Ltd, London, pp. 437–450.
Isaac, R. (2014) Responsible tourism and development in the context of peace-building. In: Wohlmuther, C. and Wintersteiner, W. (eds) *International Handbook on Tourism and Peace*. Austria, Drava, Klagenfurt, pp. 87–100.
Jamgade, S. (2021) Catalytic effect of tourism in peacebuilding: Sustainability and peace through tourism. In: Silva, J.T., Breda, Z. and Carbone, F. (eds) *Role and Impact of Tourism in Peacebuilding and Conflict Transformation. IGI Global*. Hershey, Pennsylvania, pp. 29–45.
Kaldor, M. (2003) The idea of global civil society. *International Affairs* 79(3), 583–593.
Kaldor, M. (2020) Global civil society: An answer to war. In: Seidman, S. and Alexander, J.C. (eds) *The New Social Theory Reader*, 2nd edn (eBook). Routledge, London.
Keane, J. (2003) *Global Civil Society*. Cambridge University Press, Cambridge.
McDonald, J.W. (2006) Citizen diplomacy in a changing world. *Exopolitics Journal* 2(1), 1–19.
Melissen, J. (2013) Conclusions and recommendations on public diplomacy in Europe. In: Cross, M.K.D. and Melissen, J. (eds) *European Public Diplomacy: Soft Power at Work*. Palgrave Macmillan, New York, pp. 205–212.
Mingst, K.A., McKibben, H.E. and Arreguin-Toft, I.M. (2018) *Essentials of International Relations*, 8th edn. WW Norton & Company, New York.
Moufakkir, O. (2010) Re-evaluating political tourism in the Holy Land: Towards a conceptualization of peace tourism. In: Moufakkir, O. and Kelly, I. (eds) *Tourism, Progress and Peace*. CAB International, Wallingford, UK, pp. 162–178.
Nobel Prize (2024) All Nobel Peace Prizes. Available at: https://www.nobelprize.org/prizes/lists/all-nobel-peace-prizes/2019-2010/ (accessed 10 April 2024).
Noy, C. (2013) Peace activism in tourism: Two case studies (and a few reflections) in Jerusalem. In: Blanchard, L.-A. and Higgins-Desbiolles, F. (eds) *Peace Through Tourism: Promoting Human Security Through International Citizenship*. Routledge, London, pp. 204–216.
Schneider, E. (2020) It changed my sympathy, not my opinion: Alternative Jewish tourism to the occupied Palestinian territories. *Sociological Focus* 53(4), 378–398.
Schneider, E.M. (2019) Touring for peace: The role of dual-narrative tours in creating transnational activists. *International Journal of Tourism Cities* 5(2), 200–218.
Temby, O. (2015) What are levels of analysis and what do they contribute to international relations theory? *Cambridge Review of International Affairs* 28(4), 721–742.
Tyler, M.C. and Beyerinck, C. (2016) Citizen diplomacy. In: Constantinou, C.M., Kerr, P. and Sharp, P. (eds) *The SAGE Handbook of Diplomacy*. SAGE Publications Ltd, London, pp. 521–529.
United Nations (2024) Goal 16: Promote peaceful and inclusive societies for sustainable development, provide access to justice for all and build effective, accountable and inclusive institutions at all levels. Available at: https://sdgs.un.org/goals/goal16#targets_and_indicators (accessed 11 April 2024).
Wintersteiner, W. and Wohlmuther, C. (2014) Peace sensitive tourism: How tourism can contribute to peace. In: Wohlmuther, C. and Wintersteiner, W. (eds) *International Handbook on Tourism and Peace*. Austria, Drava, Klagenfurt, pp. 31–61.
Women of the Sun (2024) Homepage on Facebook. Available at: https://www.facebook.com/groups/4012585162089853/media/?_rdr (accessed 5 April 2024).
Women Wage Peace (2024a) Action Plan. Available at: https://www.womenwagepeace.org.il/en/action-plan/ (accessed 5 April 2024).
Women Wage Peace (2024b) Events. Available at: https://www.womenwagepeace.org.il/en/category/events/ (accessed 5 April 2024).
Yaniv, L. (2013) *People-to-People Peace Making: The Role of Citizen Diplomacy in the Israeli–Palestinian Conflict*. University of Southern California Center on Public Diplomacy, Los Angeles, California.

11 The Role of Niche Tourism in Social Cohesion, Justice and Peace with a Perspective of Cross-Border Tourism

Gülsel Çiftci[1,2*]

[1]School of Community Resources and Development, Arizona State University, Phoenix, Arizona, USA; [2]Faculty of Applied Sciences, Trakya University, Edirne, Türkiye

Abstract

This chapter explores the role of cross-border tourism in fostering social cohesion, justice and peace. Historically, tourism activities have been shaped by governmental boundaries, with border cities hosting diverse forms of tourism including shopping and medical tourism. Cross-border tourism plays a pivotal role in promoting economic development, facilitating cultural exchange and engaging in participatory peacemaking efforts, thereby fostering harmony and collaboration between partitioned countries. The chapter conducts a systematic literature review to critically examine the current state of knowledge and identify research gaps concerning the impact of cross-border tourism on social cohesion, justice and peace. Following rigorous guidelines set forth by Petticrew and Roberts (2008) and Kitchenham *et al.* (2009), this review comprehensively analyses peer-reviewed journal articles to address key research questions. The findings underscore the significance of inclusive, equitable and cooperative tourism initiatives in bridging divides between communities and nations. By synthesizing existing literature, this chapter not only consolidates insights but also proposes a research agenda aimed at advancing understanding of how cross-border tourism can contribute to sustainable development and peaceful societies. Future research directions include exploring longitudinal impacts, interdisciplinary intersections with sociopolitical factors and leveraging technology for enhanced cross-border cooperation in tourism.

Keywords: Cross-border tourism, social cohesion, justice, peace, literature review

Introduction

For a very long time, human spatial and socioeconomic activities, including tourism, have been governed by the fundamental nature of government boundaries and their purposes. Economic development in all parts of the world is mainly centred on tourism because of its positive economic effects (Timothy and Gelbman, 2023). Like other destinations, border cities have hosted different forms of tourism especially shopping, medical, sports and beach tourism (Timothy and Butler, 1995). Cross-border tourism is an important part of human mobility and commerce in borderlands in all parts of the world. Country borders are important supports contributing to economic development. In many cross-border locations, shopping is a significant economic driver of tourism. Partitioned countries benefit

*Corresponding author: gulsellciftci@gmail.com

 Niche Tourism and Sustainability: Perspectives, Practices and Prospects (eds A. Farmaki *et al.*)
DOI: 10.1079/9781800626669.0011

greatly from tourism, which promotes harmony and collaboration between them. Participatory peacemaking is encouraged by tourism to promote cross-cultural understanding and collaboration (Causevic, 2010). Cross-border tourism has the potential to promote social cohesion, justice and peace by fostering intercultural exchange, promoting equitable development and building bridges between communities and nations. By embracing the principles of inclusivity, fairness and cooperation, initiatives aimed at promoting tourism across borders have the potential to contribute significantly towards encouraging shared values such as equality, peace and solidarity.

Collaboration across borders has grown in importance, especially for places where political tensions have existed for a long time but where peace is now sought after (Wolf, 2007). Peace is an integral component of sustainable development; hence, by fostering collaboration across borders via tourism, peace may be encouraged, thereby enhancing overall sustainability. Cross-border neighbouring countries are frequently compelled to work together, build political and economic ties, and make friends (Kozak and Buhalis, 2019). Encouraging inclusive, just and peaceful societies is the goal of the United Nation's (UN) Sustainable Development Goal 16 (SDG 16) to guarantee sustainable development. This has the potential to be a significant driver of sustainable tourism and raise awareness of the importance of peace via the lens of tourism (Higgins-Desbiolles *et al.*, 2022).

A few literature reviews have been conducted by researchers on cross-border tourism. From a sociopolitical perspective, Sofield (2006) contextualizes the boundaries of a state in relation to the communities along state borders as a means of investigating border tourism in border communities. In the realm of retail literature, Groß's (2015) study provides an organized classification of current knowledge on mobile shopping (m-shopping). Similarly, Choi *et al.* (2016) examine the most recent research on shopping tourism while Jin *et al.* (2017) critically evaluate previous research on tourists' shopping behaviours and offer recommendations for new study directions and methodologies. Sialverstava *et al.* (2019) identify relevant research areas from the management perspective in cross-border tourism.

Previous research has examined various aspects of tourism across borders such as economic impacts, cultural exchange, infrastructure development and environmental sustainability. The purpose of this chapter is to determine how cross-border tourism contributes to social cohesion, justice and peace. In light of this, it is necessary to review the body of prior studies. The chapter uses a novel methodology to evaluate cross-border tourism in the fields of tourism, leisure and hospitality management and proposes an agenda for future research. As such, the chapter is the first to systematically research and analyse cross-border tourism in social cohesion, justice and peace. The approach used was a systematic literature review, following the guidelines of Petticrew and Roberts (2008) and Kitchenham *et al.* (2009). This method was chosen due to its ability to comprehensively examine the existing literature and identify gaps in knowledge (Pickering *et al.*, 2015). The study aims to answer the following research questions:

1. What is the current understanding of the role of cross-border tourism in social cohesion, justice and peace in the literature?
2. Which specific topics have been addressed in the literature regarding the role of cross-border tourism in social cohesion, justice and peace?
3. What are the implications and potential areas for future research?

Theoretical Background

The phenomenon of tourism, particularly within border regions, holds significant implications for social cohesion, justice and peace. Historically, tourism has been shaped by governmental boundaries and their underlying objectives. Border cities, alongside other destinations, have hosted various forms of tourism, ranging from shopping to medical tourism (Wintersteiner and Wohlmuther, 2014; Paasi *et al.*, 2018). Cross-border tourism refers to travel activities that cross international boundaries or borders between two or more countries (Weidenfeld, 2013). It encompasses a wide range of travel experiences including leisure, recreation, business and cultural exchange, undertaken by individuals or groups moving across borders for tourism purposes. Cross-border tourism can take various forms, from short-term visits for sightseeing or relaxation to more extended stays for immersive cultural experiences, adventure travel or business engagements. It encompasses both inbound tourism where travellers visit a foreign country, and outbound tourism, which concerns residents of one country travelling to other countries for tourism purposes (Timothy and Butler, 1995). For a very long time, human spatial behaviours and socio-economic activities, including tourism, have been governed, facilitated and constrained simultaneously by political boundaries and their various functions. Almost every region on Earth has homed in on tourism as a medium of socio-economic development largely because of its positive economic effects. In common with many other destinations, borderlands have hosted various forms of tourism, including medical mobility, sports tourism, beach tourism and shopping (Hsieh and Chang, 2006; Ormond and Sulianti, 2017; Weaver *et al.*, 2023). Border regions are especially popular nature- and culture-based tourist destinations owing to their (typically) peripheral locations in the state, which often translates into relatively isolated and untouched natural environments and Indigenous cultures.

Tourism across borders promotes cultural exchange and interaction, enabling people from diverse countries to connect and learn from each other. This facilitates cross-cultural understanding, appreciation and education. Tourism generates economic benefits for destination countries, including revenue from visitor spending, job creation, infrastructure development, and the promotion of local industries and services (Honey and Gilpin, 2009). Cross-border tourism often involves collaboration and cooperation between neighbouring countries to promote tourism, develop cross-border infrastructure and facilitate travel across borders. Sustainable cross-border tourism aims to minimize negative environmental, social and cultural impacts while maximizing the positive benefits for local communities, natural resources and heritage sites (Dunets *et al.*, 2019). Cross-border tourism, in particular, plays a pivotal role in fostering economic development and collaboration between partitioned countries, thereby promoting harmony and participatory peacemaking. As emphasized by UN SDG 16, fostering inclusive, just and peaceful societies is paramount for sustainable development, with tourism serving as a potential driver in raising awareness and promoting peace through cross-cultural understanding.

The literature surrounding cross-border tourism has gradually expanded, with researchers delving into its multifaceted aspects (Stoffelen and Vanneste, 2018; Su and Li, 2021; Makkonen and Williams, 2024). However, a systematic examination of its role in social cohesion, justice and peace remains relatively scarce. The present chapter aims to address this gap by employing a novel methodology within tourism, leisure and hospitality management. By posing pertinent research questions and adhering to established guidelines for systematic literature reviews, this chapter endeavours to unravel the current understanding, identify key issues and delineate future directions for research in cross-border tourism's contribution to social cohesion, justice and peace.

Cross-border tourism is a niche form of travel that involves crossing national borders to visit destinations in neighbouring countries. This concept includes traditional aspects of tourism and contributes to promoting peace and sustainability (Hardi *et al.*, 2021). The Schengen Area enables easy travel across many European countries without border checks, which encourages cross-border tourism. Turkey shares borders with Bulgaria and Greece, and these borders are very active with cross-border tourist activities (Kozak and Buhalis, 2019; Özyurt *et al.*, 2021). Additionally, destinations such as Niagara Falls, which straddle the border between the USA and Canada, attract millions of tourists from both sides.

Cross-border tourism plays a vital role in promoting understanding and cooperation between neighbouring countries. Tourists have the opportunity to learn about different cultures, which fosters mutual respect and understanding. For instance, cross-border festivals in the European Union commemorate shared heritage (Sternberg, 2017). Moreover, tourism contributes to the economic development of border communities, creating economic ties that encourage peaceful relations. The Peace Parks in Southern Africa, such as the Great Limpopo

Transfrontier Park, are excellent examples where tourism supports conservation efforts and regional stability (Gelbman and Schweitzer, 2023). Cross-border tourism projects often require collaboration on security and infrastructure, promoting dialogue and cooperation. The Nordic Passport Union, established in 1954, is a historical example where cooperation over travel facilitated broader peace agreements among Scandinavian countries (Mårten, 2019).

Cross-border tourism plays a significant role in promoting sustainability by contributing to the management of ecosystems that cross national boundaries and fostering biodiversity, as evidenced by transboundary parks and protected areas such as the Kavango–Zambezi Transfrontier Conservation Area in Africa (Stoldt *et al.*, 2020). Moreover, collaborative investments in environmentally friendly infrastructure, including cross-border cycling routes and public transportation, play a crucial role in reducing the environmental impact of tourism and further supporting sustainability efforts (Bhattarai and Conway, 2021). Border regions often develop diversified economies through tourism, reducing dependence on single industries and increasing resilience to economic shocks. For instance, the tourism industry in the Alps benefits multiple countries, including France, Italy and Switzerland (Moreno-Gené *et al.*, 2018).

The role of tourism in peacebuilding has been explored in various contexts, such as post-conflict regions. In these regions, the reintroduction of tourism can help rebuild the economy, foster intercultural understanding and create shared spaces for dialogue and reconciliation. For example, in countries like Rwanda, which experienced intense conflict during the genocide, tourism initiatives focusing on the country's natural beauty and wildlife have not only boosted the economy but also helped reshape its international image and promote a narrative of recovery and unity. Similarly, in Northern Ireland, tourism projects such as the Peace Bridge in Derry and the development of the Titanic Quarter in Belfast have contributed to the peace process by encouraging both locals and visitors to engage with the region's history constructively. These examples demonstrate that tourism is a powerful tool for economic regeneration and for creating environments where peace and understanding can flourish (Summers, 2019; Millawithanachchi, 2020; Hadaway, 2021). By integrating these aspects, cross-border tourism not only enhances travel experiences but also contributes to broader goals of peace and sustainability (Farmaki and Stergiou, 2021).

Methods

The methodology employed in this chapter involves a rigorous selection process focusing on pertinent publications, specifically peer-reviewed journal articles from reputable English-language journals indexed in recognized databases. This approach ensures the reliability and credibility of the findings by adhering to stringent inclusion criteria and employing comprehensive search strategies. The systematic review aims to illuminate the complex interrelationships between cross-border tourism and its impacts on social dynamics, justice systems and efforts towards peacebuilding. Through meticulous selection and analysis of literature, this study endeavours to provide a thorough understanding of how cross-border tourism influences various societal aspects. By synthesizing existing research and identifying gaps in knowledge, the chapter lays a solid foundation for future investigations aimed at unravelling the intricate connections between tourism and the promotion of inclusive, just and peaceful societies, particularly in the context of cross-border interactions.

Moreover, the review sets out to critically examine the methodologies and theoretical frameworks utilized in existing studies, offering insights into both the strengths and limitations of current research. By doing so, it seeks to advance methodological rigour within the field and guide future research directions towards more comprehensive and nuanced explorations of cross-border tourism's role in societal development and peacebuilding efforts. This systematic review is based on the guidelines of Petticrew and Roberts (2008) and Kitchenham *et al.* (2009). This approach was applied because an examination of the literature can map the limits of what is now known and reveal what is still unknown (Pickering *et al.*, 2015). A review procedure was created based on the study objectives to direct the literature search, including details about the databases, search keywords and screening criteria. To identify studies on cross-border tourism, the terms 'peace + cross-border', 'peace + cross-border tourism', 'justice + cross-border', 'justice + cross-border tourism', 'social cohesion + cross-border' and 'social cohesion + cross-border tourism', were used as the search terms. According to Pickering and Byrne (2014), the keywords for a systematic literature review should identify as much pertinent material as feasible while avoiding going too far into other areas. The preliminary method for selecting appropriate publications involved scanning the titles, keywords and abstracts of articles for specific search terms. Subsequently, each study selected was evaluated against a thorough set of inclusion criteria to determine whether it was pertinent to the research topic (Ulker-Demirel and Ciftci, 2020). To uphold the review's integrity and optimize the synthesis process, only published, peer-reviewed research articles in English were considered valid. For the investigation, only the first 50 journals indexed in the Science Citation Index (SCI), Social Sciences Citation Index (SSCI) and Emerging Sources Citation Index (ESCI) were considered. Cross-border tourism research in other journals was not considered in this study. In contrast to earlier literature content reviews, it is preferable to increase both the scope of the study and its thoroughness. Using the CiteScore metrics up to 2022, 50 journals were identified. To access these journals, it utilized the Scopus academic database, as well as EBSCO, Elsevier, Proquest and Emerald.

A total of 1515 studies were screened in the tourism, leisure and hospitality management fields; 90 records, the publications that were excluded from consideration, were those that did not belong to the category of journal articles such as conference papers and book chapters. These publications were deemed irrelevant and therefore not included in the analysis as they did not meet the criteria for selection in the study. In the end, the query produced 1425 complete texts that were then further screened. It was found that 30 full-text papers were suitable after the conclusive analysis, and their reference lists were thoroughly reviewed to identify any possible absent publications.

Results

In March 2024, a literature search was conducted without a time frame, with 30 articles meeting the appropriate criteria for the study. The graph depicted in Fig. 11.1 illustrates the number of articles that have been published for each year. The largest quantity of articles was released in 2019 (n=5), followed by 2022 (n=3).

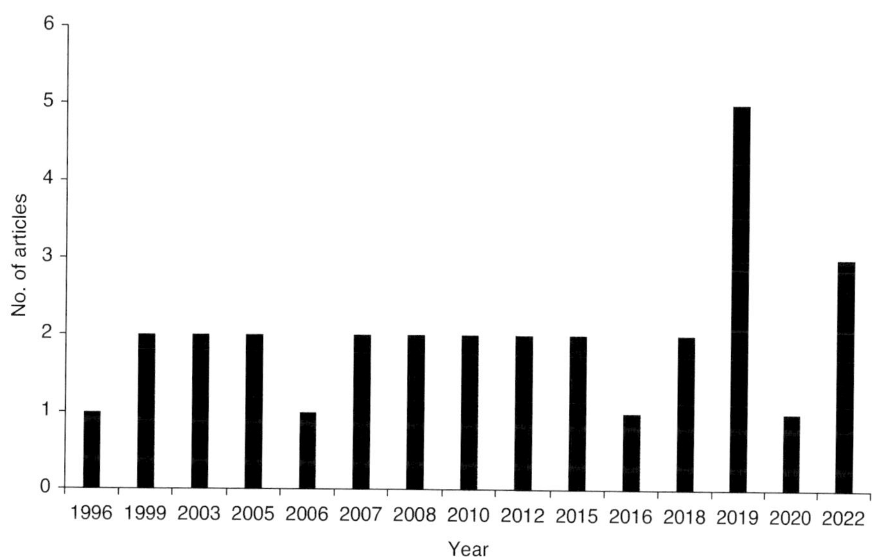

Fig. 11.1. Number of articles by year.

Table 11.1 shows the number of papers in the main journals. *Tourism Geographies* (n=5), *Journal of Sustainable Tourism* (n=4) and *Journal of Recreation Research* (n=3) published the highest number of articles on the topic. Out of the first 50 tourism, hospitality and leisure journals, only 15 included articles on cross-border tourism with a focus on social cohesion, justice and peace.

The studies examine how cross-border tourism can contribute to community peacebuilding, promote sustainable tourism and act as a mechanism to promote peace between countries. Some studies investigate the perception of international boundaries as barriers to tourism (Hannonen and Prokkola, 2023). Others consider whether sporting contact between spectators can offer the potential to promote longer-term peace (Beech *et al.*, 2005). The results also show the challenges to cross-border cooperation in tourism marketing and explore the potential for demilitarized zone peace parks in promoting peace between countries (Kim, 2022). Additionally, some studies examine the role of tourism in promoting peace around transboundary protected areas and explore intersections of peace, justice and sustainability through indigenous tourism (Higgins-Desbiolles *et al.*, 2022).

The role of tourism in peacebuilding has been explored in contexts like post-conflict regions where the reintroduction of tourism can help to rebuild the economy, foster intercultural understanding and create shared spaces for dialogue and reconciliation. For example, Anson (1999) examined the role of tourism in promoting peace in Northern Ireland after 30 years of conflict. The study found that tourism projects, such as the Peace Bridge in Derry and the Titanic Quarter in Belfast, encouraged both locals and visitors to engage with the region's history in a constructive manner. These initiatives have become symbols of unity, facilitating social interaction between historically divided communities and fostering a sense of shared space and common identity. Timothy (1999) studied cross-border partnerships in sustainable tourism between the USA and Canada, revealing that such partnerships can lead to mutual benefits and a deeper understanding between nations. This collaboration in tourism development was found to enhance environmental conservation efforts and promote peace by building trust and cooperation.

Table 11.1. Number of articles by journal.

Journal	No. of articles
Tourism Geographies	5
Journal of Sustainable Tourism	4
Tourism Recreation Research	3
Tourism Management	2
Annals of Tourism Research	2
Journal of Travel Research	2
Journal of Travel & Tourism Marketing	2
Current Issues in Tourism	2
Asia Pacific Journal of Tourism Research	2
International Journal of Contemporary Hospitality Management	1
Journal of Destination Marketing & Management	1
Tourism Review	1
International Journal of Tourism Research	1
Journal of Heritage Tourism	1
Tourism Planning & Development	1
TOTAL	30

Kim and Prideaux (2003) investigated the relationship between domestic opinion in South Korea and tourism development in North Korea. Their findings showed that positive perceptions of tourism in North Korea among South Koreans could potentially pave the way for improved inter-Korean relations and peacebuilding efforts. Timothy and Tosun (2003) analysed how international boundaries affect tourism and peace, with a focus on the USA–Canada border. They discovered that tourism can act as a bridge, promoting peace and understanding by encouraging cross-border interactions and reducing tensions associated with boundary disputes. Beech *et al.* (2005) explored the potential of sporting events in promoting long-term peace. That study highlighted that international sporting events can bring people together, creating opportunities for cultural exchange and fostering a spirit of camaraderie that transcends political and social divides. Tosun *et al.* (2005) identified strategies and challenges in cross-border tourism marketing. Their research emphasized the importance of collaborative marketing efforts to overcome political and logistical challenges, ultimately contributing to peace and cooperation between neighbouring countries. Webster and Timothy (2006) investigated why many Greek Cypriots refuse to cross the Green Line for tourism. Their study found that deep-seated historical grievances and mistrust hinder tourism's potential to promote peace. Addressing these issues requires more than tourism initiatives alone; it necessitates broader reconciliation efforts. Kim *et al.* (2007) assessed if tourism can promote peace in conflict zones. Their findings suggested that tourism has the potential to promote peace by providing economic incentives for stability and creating platforms for dialogue and cultural exchange.

Shin (2007) explored the potential of demilitarized zone peace parks in promoting peace and conservation. That study concluded that such peace parks could serve dual purposes of environmental protection and fostering peace by symbolizing reconciliation and cooperation. Gelbman (2008) analysed Israel's border tourism attractions, finding that tourism along borders can promote peace by encouraging interaction and mutual understanding between people from different sides of the conflict. Weaver (2010) discussed how tourism can promote peace and economic equity. Weaver's research highlighted that inclusive and equitable tourism development can reduce inequalities and promote social harmony, contributing to lasting peace. Gelbman and Maoz (2012) studied how the Island of Peace is introduced to Israeli tourists, finding that presenting such sites as symbols of cooperation and reconciliation can positively influence visitors' perceptions and promote peace.

Mansfeld and Korman (2015) explored conflict heritage tourism in border areas, showing that tourism can turn sites of conflict into spaces for education and reflection, thereby contributing to peace and understanding. Chung *et al.* (2016) applied the generational cohort concept to tourism and peace, demonstrating that different generational attitudes towards tourism can influence its effectiveness in promoting peace. Younger generations were found to be more open to using tourism as a tool for peacebuilding.

Chhabra (2018) suggested that tourism can foster peace relations between countries by promoting cultural exchange and mutual understanding, which are essential for long-term peacebuilding. Gelbman (2021) analysed the potential of a cross-border sport tourism event in promoting peace between Israelis and Palestinians. The study found that such events can create neutral spaces for interaction, reducing prejudices and fostering goodwill.

Trogisch and Fletcher (2022) focused on promoting peace around transboundary protected areas. Their research highlighted that collaborative management of these areas can lead to joint conservation efforts and peacebuilding through shared environmental stewardship. Mika and Scheyvens (2022) explored intersections of peace, justice and sustainability through indigenous tourism. They found that indigenous tourism initiatives can promote peace by empowering local communities, preserving cultural heritage and fostering sustainable development. Through these examples, tourism emerges as a powerful tool for not only economic regeneration but also for creating environments where peace and understanding can flourish. These findings emphasize the potential of well-planned tourism initiatives to contribute significantly to peacebuilding in post-conflict regions by supporting economic recovery, fostering intercultural dialogue and promoting a shared sense of identity and purpose.

Conclusion and Implications

In conclusion, the examination of cross-border tourism's role in social cohesion, justice and peace emphasizes its significance as a catalyst for fostering inclusive and harmonious societies. Through a systematic review of existing literature and a meticulous analysis of pertinent studies, this research has illuminated the multifaceted impacts of tourism within border regions. From promoting cross-cultural understanding to facilitating participatory peacemaking efforts, cross-border tourism emerges as a potent force in bridging divides and nurturing collaborative relationships between nations. Furthermore, this study's methodological approach, guided by established frameworks for systematic literature reviews, ensures the rigour and integrity of its findings. By synthesizing diverse perspectives and identifying key themes and research gaps, this research contributes to our understanding of cross-border tourism.

Moving forward, it is imperative to leverage the insights gleaned from this study to inform policy making and strategic planning initiatives aimed at harnessing the potential of cross-border tourism as a vehicle for sustainable development and peacebuilding. By promoting dialogue, cooperation and mutual understanding across borders, stakeholders can unlock new opportunities for economic growth, social cohesion and justice within border regions. In essence, the exploration of cross-border tourism's role in social cohesion, justice and peace emphasizes its transformative potential in shaping more inclusive and equitable societies. As we navigate the complexities of an interconnected world, embracing the transformative power of tourism offers a pathway towards realizing the shared aspirations of peace, justice and prosperity for all.

Overall, the exploration of cross-border tourism's role in social cohesion, justice and peace presents a nuanced and multifaceted landscape. This chapter, anchored in the backdrop of UN SDG 16, highlights the potential of tourism as a catalyst for fostering inclusive, just and peaceful societies. By delving into the literature spanning diverse fields such as tourism, leisure and hospitality management, this research endeavours to illuminate the intricate interplay between tourism activities and their impact on social dynamics, justice systems and peacebuilding efforts across borders.

The systematic review, adhering to established guidelines, has culminated in a comprehensive analysis of 30 pertinent articles from reputable journals. The findings reveal a growing recognition of tourism's potential to transcend political boundaries, facilitate cross-cultural understanding and nurture collaborative partnerships between nations. The studies examined within this review encompass a wide spectrum of themes, ranging from community peacebuilding initiatives to the promotion of sustainable tourism practices. They also delve into the complexities of cross-border cooperation in tourism marketing, the potential of sporting events in fostering long-term peace and the role of tourism in conflict zones. Moreover, the exploration extends to innovative concepts such as demilitarized zone peace parks and indigenous tourism, further enriching our understanding of the intersections between peace, justice and sustainability within the tourism landscape.

Each study offers unique insights and perspectives, contributing to a holistic understanding of the potential pathways through which tourism can contribute to peace and social cohesion. The literature reviewed highlights the importance of tourism as a tool for building bridges, fostering dialogue and promoting mutual understanding. It examines historical contexts such as Northern Ireland's peace process and explores contemporary challenges and opportunities in conflict-affected regions like Cyprus, Israel and Palestine, and Bosnia and Herzegovina. In essence, this study emphasizes the imperative of viewing tourism not merely as an economic activity but as a potent force for advancing social justice, fostering inclusive societies and building enduring peace across borders. As we navigate the complexities of a rapidly changing world, it is important to stay mindful of the challenges and opportunities that lie ahead. The insights gleaned from this research offer valuable guidance for policy makers, practitioners and scholars alike in harnessing the full potential of tourism as a force for positive change in the pursuit of a more just, equitable and peaceful world.

Theoretical implications

The outcomes of this study hold significant implications for the field of tourism studies and beyond. To begin with, this research sheds light on the multifaceted roles played by cross-border tourism in promoting social cohesion, justice and peace, thereby broadening our comprehension of tourism beyond its economic aspects. It emphasizes the need to acknowledge tourism as a sociopolitical phenomenon that intersects with wider issues of conflict resolution and peacebuilding. Moreover, this investigation brings attention to the intricate dynamics within border regions where tourism can be a force for unity as well as a potential source of conflict. By illuminating the influence of tourism on social dynamics and justice systems in these contexts, this research contributes to theoretical frameworks aiming to grasp the relationship between tourism, politics and society.

The analysis delves into the intersection of peace tourism and sustainability, exploring their potential to foster sustainable development and facilitate peace and justice. It emphasizes the interplay between peacebuilding and sustainable practices, introducing innovative conceptual models that leverage cross-border tourism for peace efforts. For instance, it delves into the concept of demilitarized zone peace parks, which merge environmental preservation with peacebuilding endeavours. The analysis also enriches current tourism theories by integrating insights from conflict-affected and post-conflict areas. It presents real-world evidence from case studies in various locations, such as Northern Ireland, the USA–Canada border, South and North Korea. These case studies showcase how tourism initiatives can contribute to peace by promoting intercultural understanding, establishing spaces for dialogue and aiding economic recovery in post-conflict regions. The chapter offers policy suggestions, including advocating for cross-border tourism ventures, bolstering the establishment of peace parks and endorsing community-based tourism initiatives that prioritize sustainability and justice. Overall, the review lays the groundwork for future research in tourism studies, underscoring the significance of rigorous empirical examination and considering the sociopolitical aspects of tourism.

This review propels our comprehension of the subject by shedding light on the role of tourism in peacebuilding, the intricacies and tactics involved in cross-border tourism and the potential of specialized tourism sectors. The practical implications proffer valuable guidance for policy makers and professionals seeking to leverage tourism's potential for sustainable development and peacebuilding. Furthermore, the structured methodology employed in this study serves as a model for subsequent research initiatives aiming to explore intricate topics within the field of tourism studies. The findings of this study state the significance of sustainable tourism practices in cross-border tourism, expanding theoretical frameworks beyond economic considerations. Academics who acknowledge tourism's ability to foster social unity, justice and peace contribute to a holistic understanding of sustainable tourism. This approach accentuates responsible tourism practices that prioritize the well-being of local communities, preserve cultural heritage and safeguard natural environments. Infusing sustainability principles into theoretical frameworks enriches conversations about the role of tourism in furthering not only economic growth but also social equity and environmental conservation within border regions.

Practical implications

The results of this study have significant implications for policy makers, tourism stakeholders, and practitioners involved in border regions and beyond. Firstly, policy makers can recognize the potential of cross-border tourism as a tool for fostering social cohesion and peacebuilding. They can develop targeted interventions and initiatives aimed at harnessing tourism's positive impacts, without compromising ethical considerations and community engagement.

Cross-border tourism encompasses a variety of activities that can foster peace and cooperation between neighbouring countries. Timothy (1999) studied cross-border partnerships in sustainable tourism between the USA and Canada. Such partnerships lead to mutual benefits and deeper understanding between nations, enhancing environmental conservation efforts and promoting peace by building trust and cooperation. Anson (1999) examined the role of tourism in promoting peace in Northern Ireland. Tourism projects encourage locals and visitors to engage with the region's history constructively, fostering social interaction between historically divided communities and promoting a shared sense of identity. Shin (2007) explored the potential of demilitarized zone peace parks in promoting peace and conservation. These parks serve dual purposes of environmental protection and fostering peace by symbolizing reconciliation and cooperation. Mansfeld and Korman (2015) explored conflict heritage tourism in border areas, showing that tourism can turn sites of conflict into spaces for education and reflection, thereby contributing to peace and understanding. Gelbman (2021) analysed the potential of a cross-border sport tourism event in promoting peace between Israelis and Palestinians. Such events create neutral spaces for interaction, reduce prejudices and foster goodwill through cultural exchange. Trogisch and Fletcher (2022) focused on promoting peace around transboundary protected areas. Their research highlighted that collaborative management of these areas can lead to joint conservation efforts and peacebuilding through shared environmental stewardship.

These studies demonstrate how cross-border tourism can go beyond economic benefits to foster peace, mutual understanding and cooperation between neighbouring countries. Also, tourism stakeholders, including destination managers and tour operators, can utilize the study's findings to design and implement responsible tourism practices. These practices promote intercultural dialogue, respect for local communities and sustainable development within border regions. By prioritizing ethical considerations and community engagement, tourism practitioners can contribute to building more resilient and inclusive tourism destinations.

Furthermore, this chapter highlights the importance of cross-border cooperation and collaboration in realizing the full potential of tourism as a driver of peace and justice. By fostering partnerships between neighbouring countries and investing in cross-border infrastructure and communication networks, policy makers and stakeholders can create an enabling environment for sustainable tourism development and sociopolitical reconciliation. To maximize long-term social, economic and environmental benefits, policy makers and tourism stakeholders should incorporate sustainability principles into cross-border tourism initiatives, including demilitarized zone peace parks, transboundary protected areas, cross-border cultural festivals and joint partnerships. By adopting sustainable tourism practices, destination managers can mitigate the negative impacts of tourism, such as cultural commodification and environmental degradation. They can also maximize positive outcomes for local communities, thereby enhancing the resilience and authenticity of tourism destinations. Finally, policy makers can create opportunities for inclusive growth and equitable access to tourism benefits within border regions. They can do so by prioritizing sustainability in tourism development strategies.

Recommendations for further research

This study identifies several areas for further research in cross-border tourism. Firstly, future research should investigate the long-term effects of cross-border tourism initiatives on social cohesion, justice systems and peacebuilding in specific border regions. Longitudinal research and case analyses can provide valuable insights into the sustainability and scalability of tourism interventions in conflict-affected areas, revealing how these initiatives evolve over time and their enduring impacts on local communities and international relations. Secondly, there is a need for interdisciplinary research to examine the intersectionality of tourism with other sociopolitical factors, such as migration, displacement, and identity politics within border regions. A holistic approach that considers the complex interplay of these factors can help researchers develop a more nuanced understanding of the dynamics shaping tourism development and conflict resolution. Collaborations between tourism scholars, sociologists, political scientists and anthropologists would be particularly beneficial in this regard. Thirdly, researchers should explore the role of technology and digital platforms in facilitating cross-border tourism experiences and promoting intercultural exchange. Innovative approaches to virtual tourism and digital storytelling could open new avenues for cross-border cooperation and dialogue, especially in contexts where physical travel is constrained by political or logistical barriers.

Furthermore, future research should investigate the intersection of sustainability and cross-border tourism by examining the environmental and socio-economic impacts of sustainable tourism initiatives. This includes evaluating the effectiveness and scalability of sustainability strategies in diverse cross-border contexts. Research can identify best practices for minimizing environmental footprints while maximizing socio-economic benefits for local communities. Integrating insights from environmental science, social psychology and political economy can enrich our understanding of the complex dynamics shaping sustainable tourism development within border regions and help identify conditions under which sustainable tourism can thrive and contribute to peacebuilding and social equity. Additionally, researchers should explore the potential of innovative technologies and digital platforms as tools for promoting sustainable cross-border tourism experiences. This includes investigating the use of mobile applications, social media and digital marketing to enhance tourist engagement, provide real-time information on sustainable practices and facilitate cross-border collaborations. Understanding how technology can support sustainable tourism can help stakeholders design more effective and impactful initiatives.

Finally, future studies should prioritize understanding the perspectives of local communities involved in cross-border tourism initiatives. Research should focus on how these communities perceive the benefits and challenges of such initiatives and how their participation can be enhanced to ensure equitable outcomes. Investigating how different governance structures and policies influence the success of cross-border tourism initiatives is also crucial. Identifying best practices for fostering cooperation between neighbouring countries can further enhance the effectiveness of these initiatives. By addressing these recommendations, future research can provide a comprehensive understanding of the multifaceted impacts of cross-border tourism and contribute to the development of effective, sustainable and peace-promoting tourism initiatives.

References

Anson, C. (1999) Planning for peace: The role of tourism in the aftermath of violence. *Journal of Travel Research* 38(1), 57–61.

Beech, J., Rigby, A., Talbot, I. and Thandi, S. (2005) Sport tourism as a means of reconciliation? The case of India–Pakistan cricket. *Tourism Recreation Research* 30(1), 83–91.

Bhattarai, K. and Conway, D. (2021) Impacts of economic growth, transportation, and tourism on the contemporary environment. In: *Contemporary Environmental Problems in Nepal: Geographic Perspectives. Advances in Asian Human–Environmental Research*. Springer, Cham, Switzerland, pp. 563–662.

Causevic, S. (2010) Tourism which erases borders: An introspection into Bosnia and Herzegovina. *Tourism, Progress and Peace*, pp. 48–64.

Chhabra, D. (2018) Soft power analysis in alienated borderline tourism. *Journal of Heritage Tourism* 13(4), 289–304.

Choi, M.J., Heo, C.Y. and Law, R. (2016) Progress in shopping tourism. *Journal of Travel & Tourism Marketing* 33(sup1), 1–24.

Chung, J.Y., Chen, C.C. and Lin, Y.H. (2016) Cross-strait tourism and generational cohorts. *Journal of Travel Research* 55(6), 813–826.

Dunets, A.N., Ivanova, V.N. and Poltarykhin, A.L. (2019) Cross-border tourism cooperation as a basis for sustainable development: A case study. *Entrepreneurship and Sustainability Issues* 6(4), 2207–2215.

Farmaki, A. and Stergiou, D. (2021) Peace and tourism: Bridging the gap through justice. *Peace & Change* 46(3), 286–309.

Gelbman, A. (2008) Border tourism in Israel: Conflict, peace, fear and hope. *Tourism Geographies* 10(2), 193–213.

Gelbman, A. (2021) Sport tourism and peace: Crossing the contested wall. *Tourism Geographies* 21(1), 163–189.

Gelbman, A. and Maoz, D. (2012) Island of peace or island of war: Tourist guiding. *Annals of Tourism Research* 39(1), 108–133.

Gelbman, A. and Schweitzer, R. (2023) Tourism in protected areas and transboundary parks for peace. In: Timothy, D.J. and Gelbman, A. (eds) *Routledge Handbook of Borders and Tourism*. Routledge, Abingdon, UK, pp. 366–378.

Groß, M. (2015) Mobile shopping: A classification framework and literature review. *International Journal of Retail & Distribution Management* 43(3), 221–241.

Hadaway, P. (2021) Cultural development and the politics of peacebuilding in Northern Ireland after the Good Friday agreement. Doctoral dissertation, University of Manchester.

Hannonen, O. and Prokkola, E.K. (2023) Physical access and perceived constraints: Borders as barriers to travel mobilities and tourism development. In: Timothy, D.J. and Gelbman, A. (eds) *Routledge Handbook of Borders and Tourism*. Routledge, Abingdon, UK, pp. 112–125.

Hardi, T., Kupi, M., Ocskay, G. and Szemerédi, E. (2021) Examining cross-border cultural tourism as an indicator of territorial integration across the Slovak–Hungarian border. *Sustainability* 13(13), 7225.

Higgins-Desbiolles, F., Blanchard, L.A. and Urbain, Y. (2022) Peace through tourism: Critical reflections on the intersections between peace, justice, sustainable development and tourism. *Journal of Sustainable Tourism* 30(2–3), 335–351.

Honey, M. and Gilpin, R. (2009) Tourism in the Developing World: Promoting Peace and Reducing Poverty. Washington, DC: United States Institute of Peace.

Hsieh, A.T. and Chang, J. (2006) Shopping and tourist night markets in Taiwan. *Tourism Management* 27(1), 138–145.

Jin, H., Moscardo, G. and Murphy, L. (2017) Making sense of tourist shopping research: A critical review. *Tourism Management* 62, 120–134.

Kim, E.J. (2022) *Making Peace with Nature: Ecological Encounters Along the Korean DMZ*. Duke University Press, Durham, North Carolina.

Kim, S.S. and Prideaux, B. (2003) Tourism, peace, politics and ideology: Impacts of the Mt. Gumgang tour project in the Korean Peninsula. *Tourism Management* 24(6), 675–685.

Kim, S.S., Prideaux, B. and Prideaux, J. (2007) Using tourism to promote peace on the Korean Peninsula. *Annals of Tourism Research* 34(2), 291–309.

Kitchenham, B., Pearl Brereton, O., Budgen, D., Turner, M., Bailey, J. *et al.* (2009) Systematic literature reviews in software engineering – a systematic literature review. *Information and Software Technology* 51(1), 7–15.

Kozak, M. and Buhalis, D. (2019) Cross-border tourism destination marketing: Prerequisites and critical success factors. *Journal of Destination Marketing & Management* 14, 100392.

Makkonen, T. and Williams, A.M. (2024) Cross-border tourism and innovation system failures. *Annals of Tourism Research* 105, 103735.

Mansfeld, Y. and Korman, T. (2015) Between war and peace: Conflict heritage tourism along three Israeli border areas. *Tourism Geographies* 17(3), 437–460.

Mårten, J. (2019) Patterns in Nordic cross-border region-building: An overview. In: Baldersheim, H. and Ståhlberg, K. (eds) *Nordic Region-Building in a European Perspective*. Routledge, London, pp. 23–48.

Mika, J.P. and Scheyvens, R.A. (2022) Te Awa Tupua: Peace, justice and sustainability through indigenous tourism. *Journal of Sustainable Tourism* 30(2–3), 637–657.

Millawithanachchi, K.S. (2020) Peace education and truth and reconciliation in Sri Lanka: Assessing policy effectiveness as panacea for post-conflict reconstruction. Doctoral dissertation, Howard University, Washington, DC.

Moreno-Gené, J., Sánchez-Pulido, L., Cristobal-Fransi, E. and Daries, N. (2018) The economic sustainability of snow tourism: The case of ski resorts in Austria, France, and Italy. *Sustainability* 10(9), 3012.

Ormond, M. and Sulianti, D. (2017) More than medical tourism: Lessons from Indonesia and Malaysia on South–South intra-regional medical travel. *Current Issues in Tourism* 20(1), 94–110.

Özyurt, B., Gürkan, G.Ç., Çiftçi, G. and Molla, A.C. (2021) Perceived quality, perceived value, customer satisfaction and intention to revisit in medical tourism: A research on patients from the Balkans. In: Cobanoglu, C., Gunlu Kucukaltan, E., Tuna, M., Basoda, A. and Dogan, S. (eds) *Proceedings of the Conference on Managing Tourism Across Continents*, University of South Florida (USF) M3 Publishing, p. 48. DOI: 10.5038/9781955833011.

Paasi, A., Prokkola, E.K., Saarinen, J. and Zimmerbauer, K. (eds) (2018) *Borderless Worlds for Whom? Ethics, Moralities and Mobilities*. Routledge, Abingdon, UK.

Petticrew, M. and Roberts, H. (2008) *Systematic Reviews in the Social Sciences: A Practical Guide*. Wiley, Hoboken, New Jersey.

Pickering, C. and Byrne, J. (2014) The benefits of publishing systematic quantitative literature reviews for PhD candidates and other early-career researchers. *Higher Education Research & Development* 33(3), 534–548.

Pickering, C., Grignon, J., Steven, R., Guitart, D. and Byrne, J. (2015) Publishing not perishing: How research students transition from novice to knowledgeable using systematic quantitative literature reviews. *Studies in Higher Education* 40(10), 1756–1769.

Shin, Y. (2007) A critical review of peace park tourism planning and development in the border region. *Tourism and Hospitality Planning & Development* 4(2), 111–120.

Sialverstava, S., Hanchar, A. and Jalinik, M. (2019) Current issues of cross-border tourism management – bibliometric analysis of research directions. *Studia Periegetica* 28(4), 73–85.

Sofield, T.H.B. (2006) Border tourism and border communities: An overview. *Tourism Geographies* 8(2), 102–121.

Sternberg, M. (2017) Transnational urban heritage? Constructing shared places in Polish–German border towns. *City* 21(3–4), 271–292.

Stoffelen, A. and Vanneste, D. (2018) The role of history and identity discourses in cross-border tourism destination development: A Vogtland case study. *Journal of Destination Marketing & Management* 8, 204–213.

Su, X. and Li, C. (2021) Bordering dynamics and the geopolitics of cross-border tourism between China and Myanmar. *Political Geography* 86, 102372.

Stoldt, M., Göttert, T., Mann, C. and Zeller, U. (2020) Transfrontier conservation areas and human-wildlife conflict: The case of the Namibian component of the Kavango–Zambezi (KAZA) TFCA. *Scientific Reports* 10(1), 7964.

Summers, H.P. (2019) Peace building: Providing opportunities for resolution through the built environment. Master's thesis, University of Maryland, College Park, Maryland.

Timothy, D.J. (1999) Cross-border partnership in tourism resource management: International parks along the US–Canada border. *Journal of Sustainable Tourism* 7(3–4), 182–205.

Timothy, D.J. and Tosun, C. (2003) Tourists' perceptions of the Canada–USA border as a barrier to tourism at the International Peace Garden. *Tourism Management* 24(4), 411–421.

Timothy, D.J. and Gelbman, A. (eds) (2023) Understanding borders and tourism: Complex relationships and evolving patterns. In: *Routledge Handbook of Borders and Tourism*. Routledge, Abingdon, UK, pp. 1–17.

Timothy, D.J. and Butler, R.W. (1995) Cross-border shopping: A North American perspective. *Annals of Tourism Research* 22(1), 16–34.

Tosun, C., Timothy, D.J., Parpairis, A. and Macdonald, D. (2005) Cross-border cooperation in tourism marketing growth strategies. *Journal of Travel & Tourism Marketing* 18(1), 5–23.

Trogisch, L. and Fletcher, R. (2022) Fortress tourism: Exploring dynamics of tourism, security and peace around the Virunga transboundary conservation area. *Journal of Sustainable Tourism* 30(2–3), 352–371.

Ulker-Demirel, E. and Ciftci, G. (2020) A systematic literature review of the theory of planned behavior in tourism, leisure and hospitality management research. *Journal of Hospitality and Tourism Management* 43, 209–219.

Weaver, D.B. (2010) Geopolitical dimensions of sustainable tourism. *Tourism Recreation Research* 35(1), 47–53.

Weaver, D., Moyle, B.D. and McLennan, C. (2023) Geopolitics and touristrategic borderlands: Balancing nationalism and globalisation in Hainan, China. *Tourism Management* 95, 104679.

Webster, C. and Timothy, D.J. (2006) Travelling to the 'other side': The occupied zone and Greek Cypriot views of crossing the Green line. *Tourism Geographies* 8(2), 162–181.
Weidenfeld, A. (2013) Tourism and cross border regional innovation systems. *Annals of Tourism Research* 42, 191–213.
Wintersteiner, W. and Wohlmuther, C. (2014) Peace sensitive tourism: How tourism can contribute to peace. In: Wohlmuther, C. and Wintersteiner, W. (eds) *International Handbook on Tourism and Peace*. Austria, Drava, Klagenfurt, pp. 31–61.
Wolf, A.T. (2007) Shared waters: Conflict and cooperation. *Annual Review of Environment and Resources* 32(1), 241–269.

12 Volunteer Tourism as a Sustainable Niche Tourism Form

Ângela Gonçalves*, Jéssica Ribeiro, Mariana Pereira and Bruno Barbosa de Sousa
School of Hospitality and Tourism, Polytechnic University of Cávado and Ave (IPCA), Barcelos, Portugal

Abstract
The aim of this chapter is to examine volunteer tourism as an emerging niche tourism form that has been gaining popularity in recent years and discuss its relevance to sustainability. In so doing, we draw from existing literature to analyse volunteer tourism as an emerging niche tourism form while attempting to identify the factors driving its popularity among tourists. Knowledge of tourist motivations for volunteer tourism may shed light on how the development of this niche tourism form can be aligned to the principles of sustainability. We then discuss the impacts of volunteer tourism on local communities in order to better understand the nexus between this niche tourism form and sustainability. Overall, this chapter illuminates our understanding of volunteer tourism and how it may be developed as a sustainable niche tourism form by destinations wishing to enhance their sustainability potential while minimizing the negative impacts associated with (niche) tourism development.

Keywords: Volunteer tourism, niche tourism, tourist motivation, impacts, sustainability

Introduction

As a result of globalization, new market trends have emerged focusing on niche consumer segments that seek customized products and services and more socially and environmentally responsible activities (Alves *et al.*, 2022; Santos *et al.*, 2022; Avolio *et al.*, 2024). The tourism industry is no exception to these trends. Within tourism, niche markets have been popularized in recent decades as a way to withstand competition, with destinations attempting to develop an array of innovative tourism products under the umbrella of 'niche tourism'. Niche tourism can be operated through a network of small companies (e.g. tour operators, hotels), making use of low barriers to entry and characteristics of the tourism industry, or by employing information technologies on a larger scale as a form of 'home industry' through transnational companies (Sousa *et al.*, 2022). Evidently, the relevance of niche tourism to sustainability is debatable as it depends on developmental factors that determine its ability to contribute to destinations' sustainable development. Indeed, there have been arguments that niche tourism and mass tourism are not necessarily mutually exclusive and often overlap (Ma *et al.*, 2020). In relation to this point, some niche tourism forms have been argued to represent an alternative developmental option contributing to sustainability. One such niche tourism form is volunteer tourism.

*Corresponding author: a30004@alunos.ipca.pt

DOI: 10.1079/9781800626669.0012

Volunteer tourism (also called voluntourism) is one of the fastest-growing examples of niche tourism, offering alternative travel experiences that respect the dimensions of sustainable tourism which revolve around the economy, society and the environment. It creates decommodified experiences that move the goals of tourism away from for-profit forms of escapism towards something that will place social, cultural and ecological value on local environments and communities (Pompurová *et al.*, 2018; Au-Yong-Oliveira *et al.*, 2021). In this context, volunteer tourism emerges as a niche tourism form that has attracted the attention of industry practitioners as tourist demand for volunteering has increased over the years. According to Parkes (2018), the volunteer tourism industry was valued at approximately US$2 billion per annum before the COVID-19 pandemic. 'Voluntourism' or 'volunteer tourism' is, thus, the act of travelling to a destination with the main objective of carrying out volunteer services in the social and/or environmental field, irrespective of the fact that these trips are motivated by altruism or personal interests. Nonc the less, given the potential negative impacts of volunteer tourism on the local community, this niche tourism form has been questioned concerning its ability to contribute to the sustainability of destinations. Even so, 'there is still a strong case to be made for the importance of volunteering' (Parkes, 2018, p. 94) as volunteers can play a significant role in implementing the Sustainable Development Goals outlined in the United Nation's Agenda 2030.

Therefore, the aim of this chapter is to critically discuss the volunteer tourism and sustainability nexus. In particular, we start off by explaining what volunteer tourism is and how it fits in within the scope of niche tourism development. Then, we attempt to understand the motivations of volunteer tourists that drive demand for this niche tourism form. Understanding tourist motivations can offer significant insights to destination planners wishing to develop volunteer tourism as part of niche tourism strategy. We then discuss the social and environmental impacts of volunteer tourism on local communities to better understand how volunteer tourism can actually be developed in order to contribute to sustainability. The chapter ends with a discussion of theoretical and practical implications.

Volunteer Tourism as a Niche Tourism Form

Volunteer tourism is a tourism form that combines leisure travel with volunteer work (Han *et al.*, 2020). Therefore, in order to comprehend the nature of volunteer tourism and its emergence as a niche tourism form, it is first necessary to understand what exactly volunteering is and how it is practised in the context of tourism.

The act of volunteering

Volunteering can be understood as a set of actions of social and community interest which are carried out by an individual who gives part of their time and skills to offer their services, without expecting a financial reward (Bandeira and Barbedo, 2015). According to de Oliveira Godinho (2018, p. 36), 'volunteering is a charitable activity that can be practiced anywhere, as long as the work done is unpaid and its practice is guided by certain principles. The practice of this activity, characterized by its solidarity nature, has been increasing, mainly at an international level, in response to social and environmental issues, human disasters related to terrorist attacks or situations of political dismemberment, with the main objective of the volunteer being to be part of the restructuring of a damaged environment'. As such, volunteering is an activity that enriches the individual who practises it both in terms of experiences and on a practical level since it develops their knowledge. The act of volunteering is carried out within the scope of projects, programmes and other forms of intervention at the service of individuals, families and the community, and is developed on a non-profit basis by public or private entities. This excludes actions that are isolated and sporadic or are determined by family, friendship or good neighbourly reasons (Bandeira and Barbedo, 2015).

The act of volunteering can be both formal and informal. For Lockstone-Binney and Ong (2024), formal and structured volunteering is identified as an activity that takes place in non-profit organizations and brings benefits to the community in which it takes place and to the volunteers themselves. It is carried out by individuals voluntarily and who do not receive any kind of payment or financial support. Parboteeah *et al.* (2004) stated that formal volunteering is more likely to be implemented by groups and/or organizations. In this context, several scholars pointed out that it is important to distinguish between managerial and non-managerial volunteering, as they have different characteristics (Bussell and Forbes, 2002). Managerial volunteers, for instance, carry out management tasks while non-managerial volunteers carry out more routine activities and have closer or more direct contact with the target audience of the beneficiary non-profit organizations. Informal volunteering, on the other hand, refers to offering assistance to those in need who are not friends or relatives of the volunteer.

Volunteer tourism

In relation to tourism, volunteering emerged as a popular activity leading to the development of a niche tourism form known as volunteer tourism. Butcher (2024) posited that 'voluntourism' or 'volunteer tourism' is considered to be the act of travelling to a destination with the main objective of developing volunteer services in the social and/or environmental sphere, whether these trips are motivated by altruism or personal interests. Volunteer tourism, thus, differs from other types of tourism because tourists don't just travel to get to know the place, a different culture, to be entertained or to rest, but mainly because they travel to develop voluntary actions. According to Wearing (2001), volunteer tourism is deeply related to sustainable development, benefiting not only the host community but also the volunteer tourists. Volunteer tourism seeks to have a positive impact on communities and the environment through volunteers (Röntynen and Tunkkari-Eskelinen, 2024) as there is a mutual benefit in performing this type of tourism, with the possibility of getting to know a country, a language, a new culture as well as being part of a project that aims to improve local people's lives.

Given its nature, volunteer tourism is often associated with a novel form of tourism known as solidarity tourism. However, volunteer tourism is more closely linked to conservation projects, preservation and improving the living conditions of communities whereas solidarity tourism supports the idea of intercultural understanding through dialogue. In the case of solidarity tourism, volunteering is used as a condition for getting to know the community, its culture and its identity via dialogue. Volunteer tourism, on the other hand, focuses on volunteer service and uses it as a tool to conserve, preserve and improve the living conditions of communities where there is obviously a mutual benefit between the tourist (volunteer) and the community. Hence, voluntourism is a type of tourism where tourists pay to take part in conservation and development projects in a guided way (Wearing, 2001; Guttentag, 2009; Yadav, 2024).

Generally speaking, volunteer tourists participate in activities that aim to reduce poverty among groups of populations, restore the environment or conduct research on certain social or environmental issues (Wearing, 2001). Examples of popular activities that are part of volunteer tourism include educational activities such as teaching English to children in poor countries, assisting with animal rescues, supporting orphaned children, providing medical assistance or building houses in earthquake-struck countries. Evidently, volunteer tourism includes a range of actions that are based on social and environmental initiatives. As such, it has been framed as an alternative tourism form that carries minimal risk of yielding negative impacts on the host society. Indeed, volunteer tourism has been argued to contribute to sustainable development through the exercise of 'care' as a practice underpinning volunteer tourism (Lee and Zhang, 2019) and, correspondingly, seen as a preferred niche tourism form that offers benefits to both the host community and the volunteer tourist seeking a meaningful experience. None the less, how volunteer tourism can be exercised sustainably remains to be answered, not only due to the potential negative impacts it may inflict on the host society but also because of its short-term nature. In this scope, as the achievement of sustainability through volunteer tourism depends on the volunteering practice, the motives and behaviours of volunteer tourists need to be examined before an understanding of the impacts of volunteer tourism is sought.

Motivations and Behaviours of Volunteer Tourists

Before examining the motivations driving volunteer tourism demand, it is important to establish the profile of the volunteer tourist. The majority of volunteers come from Western countries in North America and Europe and travel to impoverished destinations in Africa, Asia and South America for volunteer work (Avolio *et al.*, 2024). Specifically, Peru, Cuba, Bolivia, Ecuador, Honduras, El Salvador, Nicaragua, Brazil, Mexico and Guatemala are identified as the most visited Latin American countries for volunteer tourism while in Africa Morocco, Senegal, Tanzania and Mali are more popular, followed by China, Cambodia, Taiwan and India as main Asian volunteer tourist destinations (Caldas *et al.*, 2019). According to Caldas *et al.* (2019), the volunteer tourist tends to be young in age, under 30 years old. Around half of volunteer tourists engage in volunteering formally through an organization or institution whereas most volunteer tourists tend to be women.

Volunteer tourism can be part of a short-term project during the 'gap year' (Simpson, 2004), a 'mini-mission' (Brown and Morrison, 2003), or involve the participation in environmental research programmes (Ellis, 2003) or cultural change programmes (Lyons *et al.*, 2012). In this respect, Park *et al.* (2022) divided volunteer tourists into two distinct types: (i) volunteer tourists who devote their travel time to volunteer activities at the destination; and (ii) volunteer tourists with a vacation/leisure spirit who spend most of their time on leisure activities and a small fraction of their travel time on volunteering at the destination. Such a classification implies that volunteer tourists are driven to volunteering for various reasons (Caldas *et al.*, 2019). Although the concept of volunteering is founded on notions of altruism and personal development, always including working for a cause that volunteers believe in so that they feel they have accomplished something (Wearing, 2001; Casais and Sousa, 2020), there are several reasons behind volunteer tourist motives.

On the outset, a key motivation for volunteer tourism identified in the literature is altruism, pushing tourists towards caring for specific communities, promoting others' well-being, and sharing love and assistance (Han *et al.*, 2020; Park *et al.*, 2022). Yet, the altruism motive has been contested with scholars arguing that volunteer tourists are in fact driven to volunteering for personal goals and self-interest (Everingham, 2015). Whether volunteer tourists are motivated by personal (self-interest) or interpersonal factors (altruism), such motivations do not generally characterize other types of tourists (Lee and Yen, 2015). Other motivations, according to Han *et al.* (2019), include education, personal development and self-improvement. Indeed, volunteer tourists' commitment in the act of volunteering fosters a new form of education where they learn from their experiences and interactions with local populations, enriching their own cultures through these new experiences. Additionally, ego enhancement, self-improvement, personal status and the search of novel experiences emerge as significant motivational factors (Benson and Seibert, 2009; Weaver, 2015; Wearing *et al.*, 2017; Hernandez-Maskivker *et al.*, 2018; Timothy, 2022).

Notwithstanding, it has been argued that motives of volunteer tourists fluctuate between altruism and egoism as the line between the two is blurred (Mustonen, 2007). For instance, a vacation-focused tourist can become a volunteer-oriented tourist once engaged in the activity despite initial motives being egoistic. Indeed, various levels of voluntourism have been identified, ranging from shallow to deep, depending on the duration of volunteering, the destination, the motive driving the activity as well as the interaction with locals (Callanan and Thomas, 2005). Evidently, volunteer tourism is multifaceted and experienced varyingly, entailing that the impacts on the local community may vary as well.

Impacts of Volunteer Tourism

Volunteer tourism has been widely praised as an ideal form of tourism that is beneficial to everyone involved. Those who practise it are supposedly no longer indifferent hedonists but compassionate ambassadors of goodwill, and host communities are no longer objects of exploitation and commodification. Volunteer tourism has been positioned as the antithesis of mass tourism and all the problems normally associated with it (Caldas *et al.*, 2019); thus, it is regarded as a contributor to sustainability in destinations. As volunteer tourists engage in socially and environmentally responsible practices, volunteer tourism enables the development and promotion of good citizenship (Butcher and Smith, 2010). Volunteer tourism also fosters cultural understanding as tourists and the host community interact in a meaningful way. Moreover, volunteer tourism can help with alleviating poverty in poor destinations as it sustains the local economy. Overall, volunteer tourism allows destinations to deal with persistent social and environmental problems especially when the local community lacks resources, knowledge and skills. Unsurprisingly, volunteer tourism became the forerunner of alternative sustainable tourism (Wearing, 2001, 2002; Brown and Morrison, 2003; Stoddart and Rogerson, 2004; Callanan and Thomas, 2005; Caldas *et al.*, 2019).

At the same time, volunteer tourism has received a fair share of criticism despite the potential benefits it may offer to destinations. The criticism lies primarily on the negative impacts this form of tourism may inflict upon destinations. Guttentag (2009) argued that volunteer tourism may lead to a neglect of local people's desires and needs and may disrupt the local economy as volunteers take up the role of local workers. Indeed, flows of volunteer tourists to a destination may lead to perceptions among the community members that they should receive help rather than learn how to deal with societal and environmental issues on their own. Also, given the fact that volunteers are mostly Westerners visiting poor countries in Africa, Asia and South America, it is not surprising that a cultural divide may exist between tourists and the host community; perhaps one that represents a form of colonialism (Butcher and Smith, 2010). Volunteer tourists may, for instance, lack the language skills and cultural understanding needed to form meaningful relationships with the host community. Volunteer tourism has thus been accused of hindering community development, leading to cultural change in the host society. Instead of helping impoverished communities, volunteer tourists may in the longer term negatively affect the cultural fabric of the community in which they are volunteering.

Furthermore, it has been argued that volunteer tourists may put pressures on the destination's scarce resources especially as their number increases over time (Parkes, 2018). Volunteer tourism is becoming popular and, as such, represents a niche tourism form that is growing in demand; yet this growth is coupled with greater impacts on the host communities as the needs of this rapidly increasing number of volunteer tourists must be met. In addition, volunteer tourists are recognized as often lacking the necessary skills to complete satisfactory work (Hernandez-Maskivker *et al.*, 2018; Parkes, 2018) and may in fact cause more harm than good. For example, volunteer tourists may not know how to behave around vulnerable groups of the host community such as orphaned children despite their good intentions to help, causing psychological and emotional distress to members of the host society. This aspect coupled with the fact that volunteer tourism is undertaken for a short period of time raises questions over its sustainability in the future (Lee and Zhang, 2019). As one group of volunteer tourists leave the host community another one arrives, creating a burden on the host community which needs to continuously adapt to new volunteering groups. Thus, the benefits of volunteer tourism are short-term and as most studies are focused on the perspectives of volunteer tourists (Avolio *et al.*, 2024), it remains unclear as to what the host community's views and experiences of volunteer tourism are.

The Sustainability of Volunteer Tourism

In light of the criticism of volunteer tourism, it is worth mentioning that if volunteer tourism is planned and implemented correctly, it can also bring benefits to all those involved in these practices (Lee and Zhang, 2019). For volunteer tourism to overcome its shortcomings, it is important to first of all address the needs of all the stakeholders involved and ensure their perspectives are heard. Volunteer tourism stakeholders involve primarily the volunteers, the organizations and/or tour operators planning the trips, and the host community itself. The needs of the host community are particularly important, otherwise volunteer tourism development ignoring host community voices may run the risk of fulfilling the needs of volunteer tourists only. This requires research into the needs and desires of all the parties involved to ensure that developmental attempts are aligned to stakeholder interests. Research should also be done at regular intervals to measure impacts in order to take corrective action if necessary and when problems are detected.

Moreover, volunteer tourists need to be properly educated and trained in order to ensure maximum benefits to the host community are delivered and the risk of harm is minimized. According to Parkes (2018), it is important to set and follow international standards for community development and these revolve around aspects such as participatory planning, community engagement, organizing for change, promotion of inclusivity and diversity as well as policy improvement. In this context, tour operators promoting volunteer tourism programmes must be accredited to ensure that understanding on the sensitivities of volunteering is present. Likewise, volunteer tourists need to be sufficiently supervised whereas their skillsets need to be properly matched to the required activities at the destination if the full range of benefits of volunteer tourism is to be unleashed.

Notwithstanding, the goal of volunteering needs to be redefined to avoid the risk of the cycle of dependency on foreign aid. Rather than helping the host community, volunteers may teach the members of the host society the necessary skills to sustain the local economy. For example, instead of teaching English to children at the host destination volunteer tourists may teach the language to older people who can then become teachers themselves. Overall, the benefits offered by volunteer tourism need to be experienced long-term for this niche tourism form to become sustainable and such a goal demands commitment from the parties involved. Lee and Zhang (2019) posited that all three pillars of sustainability can be strengthened in terms of volunteer tourism with the input of 'care', which in turn ensures that volunteer tourism contributes to a more resilient future for both the host community and the volunteers involved.

Conclusion and Implications

Tourism is progressively characterized by smaller market segments with different individual interests (Sousa *et al.*, 2022) that came to be known as niche tourism. Niche tourism includes various forms on the basis of the purpose of the visit. As tourist needs changed over time, more socially and environmentally responsible activities became prominent, leading to the development of alternative tourism products gaining traction in global tourism. One such popular tourism form is volunteer tourism. Volunteer tourism is argued to offer several benefits to both the host community and the tourists themselves. None the less, it has also been accused of inflicting negative impacts on the host society if not managed properly.

This chapter aimed at examining volunteer tourism as an emerging niche tourism form and discuss its relevance to sustainability. While several motives appear to drive demand for volunteer tourism, the line between altruistic and egoistic motives is blurred (Mustonen, 2007). Understanding volunteer tourism motivations is important for destination planners and policy makers as well as industry practitioners including tour operators offering volunteer tourism products, as it enables the efficient development of relevant initiatives. Thus, maximum satisfaction can be achieved among volunteer tourists. At the same time, volunteer tourism has been argued to offer both positive and negative impacts at the host community. Therefore, it is necessary to ensure that development of volunteer tourism is such that minimum damage is inflicted on the host society. Otherwise, volunteer tourism may not be sustainable long-term.

Overall, this chapter illuminates our understanding of volunteer tourism and how it may be developed as a sustainable niche tourism form by destinations wishing to enhance their sustainability potential while minimizing the negative impacts associated with (niche) tourism development. This type of tourism was already a trend in pre-pandemic years, and everything indicates that its popularity will increase in the near future as war situations experienced currently in Ukraine and Palestine only have reinforced the idea of helping others (Colpin *et al.*, 2024). This chapter sought to provide some recommendations to the short-term nature of volunteer tourism and, hence, overcome any potential problems it may be associated with. Volunteer tourism needs to become more sustainable if it is to contribute to the sustainability of a destination and in doing so it must address the economic, social and environmental aspects of the concept of sustainability. Given that research on the volunteer tourism and sustainability interface is still lacking, this chapter opens up the dialogue on this important topic and aspires to contribute to the understanding of how this niche tourism form can enable destinations achieve a more sustainable tourism industry.

From this chapter it can be concluded that, if there is a will, everyone has something within themselves that can help minimize inequalities and the pain caused by them. Although several countries are far behind in volunteer tourism development (especially in Europe), it is possible that more and more destinations may embrace this niche tourism form in order to capitalize on the benefits it may offer. Such initiatives may go beyond activities focused on poverty alleviation in impoverished countries or environmental conservation and include volunteering activities in developed countries experiencing sociopolitical tensions or environmental problems, as well as encompassing vulnerable societal groups that are in need of help. Given the interdisciplinary nature of volunteer tourism, it is required that future studies are conducted to allow for a greater understanding on the nature of volunteer tourism as a niche tourism form that contributes to destination sustainability. Such research attempts need to include qualitative and quantitative data collection methods in developing and developed nations while evaluating and comparing the perspectives of the volunteer tourists, the volunteer organizations, the destination planners and of course the host community.

References

Alves, H.M., Sousa, B., Carvalho, A., Santos, V., Lopes Dias, A. *et al.* (2022) Encouraging brand attachment on consumer behaviour: Pet-friendly tourism segment. *Journal of Tourism, Heritage & Services Marketing* 8(2), 16–24.

Au-Yong-Oliveira, M., Perfeito, S. and Almeida, F. (2021) Inovação e turismo voluntário: Uma experiência aceite por todos. *Journal of Tourism and Development* 36(2), 319–328. DOI: 10.34624/rtd.v36i2.10491.

Avolio, B., Prados-Peña, M.B. and Pardo, E. (2024) Volunteer tourism: A systematic literature review. *SAGE Open* 14(2), 21582440241260924.

Bandeira, A. and Barbodo, P. (2015) O voluntariado: um recurso em valorizaçãO na sociedade portuguesa. In: *Congresso Internacional de Contabilidade e Auditoria*. Ordem dos Técnicos Oficiais de Contas e Instituto Superior de Contabilidade e Administração de Coimbra, Coimbra, Portugal.

Benson, A. and Seibert, N. (2009) Volunteer tourism: Motivations of German participants in South Africa. *Annals of Leisure Research* 12(3–4), 295–314.

Brown, S. and Morrison, A.M. (2003) Expanding volunteer vacation participation: An exploratory study on the mini-mission concept. *Tourism Recreation Research* 28(3), 73–82.

Bussell, H. and Forbes, D. (2002) Understanding the volunteer market: The what, where, who and why of volunteering. *International Journal of Nonprofit and Voluntary Sector Marketing* 7(3), 244–257.

Butcher, J. (2024) Volunteer tourism in the context of development thinking. *Tourist Studies* 24(2), 152–171.

Butcher, J. and Smith, P. (2010) "Making a difference': Volunteer tourism and development. *Tourism Recreation Research* 35(1), 27–36.

Caldas, M.I., Machado, H., Sousa, B.B. and Vareiro, L.C. (2019) Um olhar exploratório sobre o turismo voluntário. *European Journal of Applied Business and Management* 6(1), 15–30.

Callanan, M. and Thomas, S. (2005) Volunteer tourism: Deconstructing volunteer activities within a dynamic environment. In: Novelli, M. (ed.) *Niche Tourism: Contemporary Issues, Trends and Cases*. Elsevier Butterworth-Heinemann, Oxford, pp. 183–200.

Casais, B. and Sousa, B. (2020) Heterogeneity of motivations and conflicts in pilgrim-to-pilgrim interaction: A research on the way of Saint James. *Tourism Management Perspectives* 36, 100748.
Colpin, N., Dury, S. and Jegers, M. (2024) Voluntourism placements: Voluntourists' experiences and learning potential. In: *Current Issues in Tourism*. DOI: 10.1080/13683500.2024.2342402.
de Oliveira Godinho, C.I. (2018) Capacitação de comunidades locais para a gestão de modelos de negócio turísticos através do turismo voluntário: Projeto aplicado. In: *Dissertação de Mestrado, Universidade de Aveiro*. Aveiro, Portugal.
Ellis, C. (2003) Participatory environmental research in tourism: A global view. *Tourism Recreation Research* 28(3), 45–55.
Everingham, P. (2015) Intercultural exchange and mutuality in volunteer tourism: The case of Intercambio in Ecuador. *Tourist Studies* 15(2), 175–190.
Guttentag, D.A. (2009) The possible negative impacts of volunteer tourism. *International Journal of Tourism Research* 11(6), 537–551.
Han, H., Meng, B., Chua, B.L., Ryu, H.B. and Kim, W. (2019) International volunteer tourism and youth travelers – an emerging tourism trend. *Journal of Travel & Tourism Marketing* 36(5), 549–562.
Han, H., Lee, S. and Hyun, S.S. (2020) Tourism and altruistic intention: Volunteer tourism development and self-interested value. *Sustainability* 12(5), 2152.
Hernandez-Maskivker, G., Lapointe, D. and Aquino, R. (2018) The impact of volunteer tourism on local communities: A managerial perspective. *International Journal of Tourism Research* 20(5), 650–659.
Lee, H.Y. and Zhang, J.J. (2019) Rethinking sustainability in volunteer tourism. *Current Issues in Tourism* 23(14), 1820–1832.
Lee, S. and Yen, C.-L (Alan). (2015) Volunteer tourists' motivation change and intended participation. *Asia Pacific Journal of Tourism Research* 20(4), 359–377.
Lockstone-Binney, L. and Ong, F. (2024) The sustainable development goals: The contribution of tourism volunteering. In: Mooney, S., Robinson, R.N.S., Solnet, D.J. and Baum, T. (eds) *A Sustainable Tourism Workforce: Current Issues*. Routledge, London, pp. 279–296.
Lyons, K., Hanley, J., Wearing, S. and Neil, J. (2012) Gap year volunteer tourism. *Annals of Tourism Research* 39(1), 361–378.
Ma, S. (David), Kirilenko, A.P. and Stepchenkova, S. (2020) Special interest tourism is not so special after all: Big data evidence from the 2017 great American solar eclipse. *Tourism Management* 77, 104021.
Mustonen, P. (2007) Volunteer tourism – altruism or mere tourism. *Anatolia* 18(1), 97–115.
Parboteeah, K.P., Cullen, J.B. and Lim, L. (2004) Formal volunteering: A cross-national test. *Journal of World Business* 39(4), 431–441.
Park, C., Lee, S., Lee, C.K. and Reisinger, Y. (2022) Volunteer tourists' environmentally friendly behavior and support for sustainable tourism development using value–belief–norm theory: Moderating role of altruism. *Journal of Destination Marketing & Management* 25, 100712.
Parkes, E. (2018) Rethinking the true impact of voluntourism in line with new community development standards. *Whanake: The Pacific Journal of Community Development* 4(2), 90–101.
Pompurová, K., Marčeková, R., Šebová, Ľ., Sokolová, J. and Žofaj, M. (2018) Volunteer tourism as a sustainable form of tourism – the case of organized events. *Sustainability* 10(5), 1468.
Röntynen, R. and Tunkkari-Eskelinen, M. (2024) Voluntourism in the context of community-based tourism, and regenerative tourism: A theoretical exploration focusing on responsibility. In: Balkaran, R. and Hattingh, C. (eds) *Proceedings of the 7th International Conference on Tourism Research*. Academic Conferences and Publishing International Limited, Reading, pp. 340–348.
Santos, V., Sousa, B., Ramos, P. and Valeri, M. (2022) Emotions and involvement in tourism settings. *Current Issues in Tourism* 25(10), 1526–1531.
Simpson, K. (2004) 'Doing development': The gap year, volunteer-tourists and a popular practice of development. *Journal of International Development* 16, 681–692.
Sousa, B., Santos, V. and Ramos, P. (2022) Special forms of tourism. In: Buhalis, D. (ed.) *Encyclopedia of Tourism Management and Marketing*. Edward Elgar Publishing, Cheltenham, UK, pp. 199–201.
Stoddart, H. and Rogerson, C.M. (2004) Volunteer tourism: The case of habitat for humanity South Africa. *GeoJournal* 60(3), 311–318.
Timothy, D.J. (2022) Volunteer tourism, poverty, and people on the margins: Altruism, ego-enhancement, and voyeurism in the Himalaya. In: Nyaupane, G.P. and Timothy, D.J. (eds) *Tourism and Development in the Himalaya: Social, Environmental, and Economic Forces*. Routledge, Abingdon, UK, pp. 170–183.

Wearing, S. (2001) *Volunteer Tourism: Experiences That Make a Difference*. CAB International, Wallingford, UK.

Wearing, S. (2002) Re-centering the self in volunteer tourism. In: Dann, G.M.S. (ed.) *The Tourist as a Metaphor of the Social World*. CAB International, Wallingford, UK, pp. 237–262.

Wearing, S., Young, T. and Everingham, P. (2017) Evaluating volunteer tourism: Has it made a difference? *Tourism Recreation Research* 42(4), 512–521.

Weaver, D. (2015) Volunteer tourism and beyond: Motivations and barriers to participation in protected area enhancement. *Journal of Sustainable Tourism* 23(5), 683–705.

Yadav, S.J. (2024) Analyzing the influence of sustainable voluntourism on community development and empowerment. In: Jermsittiparsert, K. and Suanpang, P. (eds) *Special Interest Trends for Sustainable Tourism*. IGI Global, Hershey, Pennsylvania, pp. 152–165.

13 Conclusion

Anna Farmaki[1]* and Pramendra Singh[2]
[1]*Cyprus University of Technology, Limassol, Cyprus;* [2]*Lovely Professional University, Punjab, India*

The merits of niche tourism have long been discussed. Niche tourism may enable destinations to diversify their product offering, enhance their image and improve their competitiveness. Niche tourism is also widely regarded as an alternative to mass tourism which may generate more positive impacts on the environment and local communities. Likewise, niche tourism may provide unique experiences to tourists who are looking for more meaningful encounters during their travels. Indeed, the new-age tourist is more demanding and opts for destinations which offer novelty. An increasing number of destinations began to use their untapped cultural and environmental resources as niche tourism products to cater to the needs and preferences of this emerging tourist market. Technology and social media have further enhanced the development of niche tourism, as information about novel destinations and tourism activities is readily available. Unsurprisingly, niche tourism has grown in popularity as a tourism development strategy.

Despite the benefits that niche tourism may offer to destinations and tourists alike, it has also attracted a fair share of criticism. The criticism lies primarily on the overlap of niche tourism with mass tourism aspects as well as the potential negative impacts it may exert on destinations. Additionally, the lack of local community participation, limited knowledge and resources, regulatory aspects and environmental issues present as obstacles hindering the sustainable development of niche tourism. In this context, the relationship between niche tourism and sustainability has emerged as an important topic of discussion. None the less, the niche tourism and sustainability nexus remains an underexplored research area.

This book attempted to fill this gap in research and advance understanding of the contribution of niche tourism to sustainability. Through a collection of chapters discussing various forms of niche tourism including rural tourism, ecotourism, peace tourism and wine tourism among others, this edited book identified demand- and supply-related aspects that shape perceptions, behaviours and experiences of niche tourism. In addition, the book considered a range of practices undertaken by tourism providers in implementing niche tourism. Last, the book analysed the prospects of niche tourism forms in yielding positive impacts to local communities and the environment and, as such, contributing to sustainability in destinations. In so doing, this edited book offers a holistic perspective on niche tourism and addresses important aspects related to the niche tourism

*Corresponding author: anna.farmaki@cut.ac.cy

 Niche Tourism and Sustainability: Perspectives, Practices and Prospects (eds A. Farmaki *et al.*)
DOI: 10.1079/9781800626669.0013

and sustainability interface that may offer valuable insights to researchers and destination planners.

As a concluding remark of the book *Niche Tourism and Sustainability: Perspectives, Practices and Prospects*, we provide a summary of the main issues raised in the book's chapters in an attempt to discuss the implications arising from each case presented. In particular, recommendations are made in regard to the different niche tourism forms discussed in the chapters that may help destination planners and business managers in aligning their practices to sustainability dimensions. Suggestions for further research are also made to scholars interested in investigating niche tourism. Divided into four parts, this edited book sheds lights on important aspects that relate to niche tourism and sustainability such as tourist behaviour and experiences, resident perspectives, implementation practices and prospects of niche tourism.

In the first part of the book, three chapters discussed tourist behaviour and experiences in the context of niche tourism. Chapter 1, authored by Beverly Ntshabeleng and Madiseng Phori, explored South Africa's transition from mass tourism to sustainable niche tourism and discussed the importance of the mindful tourist in this context. The authors offer recommendations that reflect tourists' evolving preferences, which position South African tourism as sustainable through the creation of meaningful tourist experiences. In Chapter 2, Feras Orekat, Glenn Murray and Andrew Le draw from Sydney's Quarantine Station to highlight the importance of visitor experiences in heritage tourism development. The authors conclude with recommendations on the conservation of the site's cultural characteristics reflecting Australian Aboriginal culture. Werner Gronau, Jana Große Hokamp and Knut Scherhag authored Chapter 3 which discussed the attitudes and motivations of motorcycle tourists. The chapter underlines the requirements for destinations in developing motorcycle tourism in a more sustainable way.

Part II of the book examined residents' perspectives of niche tourism. Two chapters are presented in this part. Chapter 4, authored by Fani Efthymiadou and Chryso Panayidou, discussed the perceptions and support of residents of the island of Tilos in Greece for environmental practices as part of ecotourism. The chapter focused on a zero waste programme implemented on the island to conclude that residents perceive the initiative favourably and express support accordingly. As such, the chapter illustrates that similar environmental practices can complement ecotourism development attempts by promoting an appropriate image and directing tourist behaviour towards sustainability. In Chapter 5, Emrullah Erul and Furkan Atasoy Karacabey investigated the impact of Manavgat (Türkiye) residents' attachment to their community and perceptions of mass tourism on their attitudes towards niche tourism development. Strong community ties and negative perceptions of mass tourism significantly enhanced favourable attitudes and support towards niche tourism. The chapter highlights the importance of involving residents in sustainable tourism practices.

In Part III of the book, four chapters are included that discussed niche tourism implementation practices. In Chapter 6, Athina Nella and Dimitrios Stergiou analysed UK stakeholder practices in terms of birdwatching tourism and reveal varied sustainability efforts. The chapter concludes with recommendations of multifaceted strategies, including active participation from local communities, robust conservation laws, environmental education and leveraging technology, at addressing the challenges present. Chapter 7, authored by Stella Kladou, Maria Alebaki and Kleio Arvanakou, draws from the island of Corfu in Greece to explore sustainability practices and challenges for family wineries in the context of wine tourism. Analysis revealed the importance of not only sensing but also of seizing and of configuring opportunities and challenges in order to achieve sustainable wine tourism development. In Chapter 8, Sıla Karacaoğlu examined the role of creative tourism in the sustainable development of rural areas. By presenting cases of creative tourism practices in Türkiye, the author suggests that creative tourism can represent an important niche tourism form that enhances the tourist appeal of rural destinations while contributing to sustainability, as it does not require large investments and utilizes existing cultural and natural elements. Maria Hadjielia Drotarova and Vasilis Papavasiliou are the authors of Chapter 9, which explored the practices of rural small and medium tourism enterprises (SMTEs) in Cyprus.

The chapter concludes with an emphasis on digital transformation as an innovative way to deal with the challenges encountered by rural SMTEs, which may in turn unlock the potential to sustainable rural tourism development.

In Part IV of the book, the prospects of niche tourism in terms of sustainability are discussed through three chapters. In Chapter 10, Katerina Antoniou examined the role of citizen diplomacy undertaken by residents and tourists as part of peace tourism in destinations' sustainable development using the Israel/Palestine conflict as a case study. Gülsel Çiftci authored Chapter 11 in which she discussed the impact of cross-border tourism on social cohesion, justice and peace. The author highlighted how cross-border tourism can contribute to sustainable development of destinations and concludes the chapter with the provision of a research agenda. Last, in Chapter 12 authored by Ângela Gonçalves, Jéssica Ribeiro, Mariana Pereira and Bruno Barbosa de Sousa, a discussion of volunteer tourism as a sustainable niche tourism form is provided. The chapter concludes that volunteer tourism may be developed to enhance destinations' sustainability potential while minimizing the negative impacts associated with (niche) tourism development.

Overall, this book provides a collection of conceptual and empirical chapters that discuss the trends surrounding niche tourism development, the challenges and potential of niche tourism in eliciting sustainability, as well as the practices and strategies that may be adopted in strengthening the sustainability goal. It is worth mentioning that every destination is different and unique in characteristics and developmental level. Hence, the recommendations offered in this book and the niche tourism forms presented may not apply to all destinations. However, destination stakeholders and researchers may find the cases discussed in this book insightful. As demand for niche tourism is expected to increase post COVID-19, given tourists' preference for smaller travel groups and more customized experiences, the future of niche tourism seems bright. Evidently, more academic attention is needed on the ability of niche tourism to contribute to sustainability. This book makes a small contribution to the emerging discourse on niche tourism. While each chapter presents a specific niche tourism form and approaches the niche tourism and sustainability relationship from a certain angle, collectively the book offers a comprehensive overview of this important topic. We hope *Niche Tourism and Sustainability: Perspectives, Practices and Prospects* offers a stepping stone for further investigation of significant issues related to the interface between niche tourism and sustainability and, as such, provides suggestions for a more effective sustainable destination management and practice.

Index

Note: Page numbers in **bold** type refer to **figures**
Page numbers in *italic* type refer to *tables*

Printed and bound by CPI Group (UK) Ltd, Croydon, CR0 4YY

13/09/2026

14961116-0007